"*Teaching Struggling Students in Math* is clear, informative, and concise with concrete examples, ideas, and strategies for teaching rigorous and relevant mathematics instruction...Wish I had this book my first year as an 8th-grade mathematics teacher."—**Karlene Lee**, vice president ALAS-CELT; former assistant superintendent, Clark County School District, Las Vegas, Nevada

"As a former elementary teacher and principal and someone who enjoyed math as a student and took math courses through calculus, I found the book to be very informative. The philosophy on math instruction is well-structured and sensible. The concept of linkage, I believe, is an aha moment for teachers. I shared one example from the book with grade 3–5 math teachers and a light went on! I can't wait to see our teachers implement these strategies over the next couple of years."—**Laura Hirsch**, assistant superintendent, Crete-Monee School District, Crete, Illinois

"Readers of Bill's book, along with those hosting Mr. Hanlon for professional development, will readily recognize Bill's unique ability to refine scientifically research-based best practices as he effectively applies each strategy to organize and focus staff and students on the goal of measurable growth.—**Mike Rumsey**, high school principal, Okawville Jr/Sr High School, Okawville, Illinois

TEACHING STRUGGLING STUDENTS IN MATH

Too Many Grades of D or F?

Bill Hanlon

Rowman & Littlefield Education

A division of

ROWMAN & LITTLEFIELD PUBLISHERS, INC.

Lanham • New York • Toronto • Plymouth, UK

Published by Rowman & Littlefield Education
A division of Rowman & Littlefield Publishers, Inc.
A wholly owned subsidiary of The Rowman & Littlefield Publishing Group, Inc.
4501 Forbes Boulevard, Suite 200, Lanham, Maryland 20706
www.rowman.com

10 Thornbury Road, Plymouth PL6 7PP, United Kingdom

British Library Cataloguing in Publication Information Available

Library of Congress Cataloging-in-Publication Data Available
ISBN: 978-1-4758-0068-5 (cloth : alk. paper) - 978-1-4758-0069-2 (pbk : alk. paper) - 978-1-4758-0070-8 (electronic)

∞™ The paper used in this publication meets the minimum requirements of American National Standard for Information Sciences—Permanence of Paper for Printed Library Materials, ANSI/NISO Z39.48-1992.

Printed in the United States of America

CONTENTS

PREFACE

The United States' ranking in mathematics achievement falls far below expectations when compared to the achievement of other nations. Math is important. It is more than just a body of knowledge to be memorized. It is a way of thinking that affects the way we live individually and is important to the continued prosperity of our nation.

To address this national issue, most states have adopted the new, more rigorous common core state standards in mathematics. With a failure rate that currently runs about 50 percent for students enrolling in algebra, many in the math community and school administrators are bewildered. They ask how, with that high of a failure rate, we can expect *all* kids to enroll in algebra and pass with even higher standards.

Too many of our students simply don't experience success in mathematics. They struggle in math almost every year and can't wait until they don't have to take another math class. They don't see how the math they learn today is connected to the math they learned last year or how it will be connected to the math that will be taught next year. They believe the math taught in the classrooms today has no connection to how they live life.

The recommendations in this book directly address the needs of struggling students in math: students who have not experienced success in math, students living in poverty, students earning grades of D and F, and those who see little connection to life outside of school. This book answers the fundamental question that should be asked of all teachers and administrators: *What are you doing to help my child learn math?*

You should quickly notice in this book that there are two basic standards: the "My Kid" standard and the commonsense standard. The My Kid standard simply means that we should treat students in our schools the same way we would like to have our own children treated. The commonsense standard suggests that the recommendations made should appeal to educators' common sense and experiences and be backed up by research. While there are many factors that impact educational achievement, I will clearly focus on seven key recommendations that have the greatest impact on achievement and that are in our control.

Kids get only one chance at receiving a good education. We owe it to them to make sure we are doing everything possible to ensure they receive it. The content teachers teach clearly impacts student achievement, but just as importantly, so does *how* that content is taught. We know *what works is work*, but "working smart" will increase the success rate of our students. We need to keep in mind that the only time *success* comes before *work* is in the dictionary. There are no silver bullets.

Students can't be critical thinkers or problem solvers if they do not have a body of knowledge from which to draw. Too many of today's math students are not mastering or connecting the concepts and skills they are being taught. They learn too much math in isolation. Students should be taught how math is linked: for example, how the Pythagorean Theorem, the distance formula, the equation of a circle, and the trig identity $\cos^2(x) + \sin^2(x) = 1$ are all the same formula, just written differently because they are being used in different contexts.

By introducing concepts and skills in familiar language and linking those concepts and skills to previously learned math and outside experiences, we have the opportunity to review, reinforce, and address student deficiencies and make students more comfortable in their knowledge, understanding, and application of mathematics.

To create interest and enthusiasm in mathematics, students should know how the math they are learning is used outside the classroom. Besides finding the vertex, focus, and directrix of a parabola, students should realize that paraboloids are used in making headlights, flashlights, satellite dishes, and amphitheaters. They should understand that changing the tire size (circumference) on a car also changes the car's speedometer and odometer readings. Yes, we use the classroom math we learn in life every day.

A basic axiom of math is: *The more math you know, the easier math is.* All too often, students who face the most difficulty in math are the ones who don't realize there might be a better way to compute, solve, or graph an equation than the method they are employing. The secret to success in math is having a body of knowledge that is committed to memory and then making decisions based on the problem presented.

As an example, if students who are experiencing difficulty in math are asked to compute $4 \times 13 \times 25$, many of them will take out a pencil and paper and begin multiplying 13×4, then multiply that result by 25. That takes time. Students who were taught math in a more meaningful way would examine the problem for a few seconds and realize that they can use the associative and commutative properties; in their heads, they multiply 4×25, then mentally multiply that by 13, getting a result of 1,300 in seconds without writing anything down.

When the most successful students are asked to add fractions, the first thing they do is find a common denominator. Typically, they use one of four methods to find a common denominator. If the denominators are relatively prime, they would probably multiply them. If the numbers are familiar and a little larger, they might write multiples to find the common denominator. A third method is to find the least common multiple by writing the denominators as a product of primes using a factor tree. And a fourth method, if the numbers are larger and students didn't know the multiples mentally, would be the reducing method.

The point is that the computation can be made easier by choosing the most appropriate method of finding a denominator based on the denominators presented in the problem. Students who experience difficulty, who struggle, probably know only one method: multiplying the denominators. That makes math more cumbersome.

Good students know that math can be made very easy by having a body of information that allows them to make decisions or choices that makes their work easier. Students solving systems of equations know they can choose graphing, substitution, linear combination, or matrices. Successful students look at those problems, then make a decision based on the coefficients. The same decision-making process occurs all throughout mathematics. Students can solve quadratic equations using the zero product property, completing the square, or the quadratic formula. Making a good decision makes math a whole lot easier.

There are no shortcuts to increasing student achievement. Classroom teachers and administrators must work to achieve their goals. That work should be reflected in the teachers' disposition, content knowledge, preparation, and instructional and assessment practices and in how well they work with their students. If instruction is to improve, supervisors must develop expectations before the instructional year begins and give timely and specific feedback to classroom teachers on how those expectations are being implemented.

The goal of the recommendations embodied in this book is to make students more comfortable in their knowledge, understanding, and application of math by helping them to organize their learning. Preparation, instruction, student notes, homework assignments, test preparation, and assessments are all connected and focused on student expectations so the students can study more effectively and efficiently, resulting in increased classroom performance and student achievement. Emphasis is also placed on the great importance of student–teacher relationships and how those relationships impact student performance and achievement.

By implementing these highly effective, job-embedded, no-nonsense, commonsense instructional and assessment strategies today, we will increase student achievement in mathematics, allowing our students to lead the world in mathematics achievement.

ACKNOWLEDGMENT

I have been with my wife Sharon longer than I have been involved in education. She has stood by me through all the great times and accompanied or assisted me in the many endeavors in which I took part.

When I first began teaching and planned my lessons, I would discuss how I was going to teach the lesson with her so she could comment on her understanding. My belief was that if my instruction did not make sense to her, then it probably would not make sense to my students, either. She was a great sounding board—sometimes too great. On occasion, I thought I had a great lesson planned and she would pan it.

Through my teaching career, I took on added activities, such as being the department chairperson, as well as advising the National Honor Society, Key Club, and Student Council. These are clubs and organizations that typically take time beyond the school day and into the summer. Imagine how much fun we had together selling fireworks as a fundraiser in Las Vegas in late June and July. Whether I was chaperoning dances or trips or preparing assemblies or end-of-year awards nights, my wife was always at my side and actively helped to ensure that the events were successful and the students had a positive experience.

As I began speaking at national conventions, my wife was there and provided a very frank critique of my presentations. The videotapes I made and put on my website also had to have her approval. Since she is not a math person, I liked to have her critiquing the math lesson, knowing that if it wasn't clear to her, it probably would not be clear to viewers. But like any good spouse, she didn't just stop with the lesson critique; my hair, shirt, ties, eye contact, body language, facial expressions, and voice were all fair game. That kind of critique might bother some people; it did me, too.

Today, she is still involved in handling requests and calendaring professional development, conferences, speaking engagements, travel arrangements, and anything else that comes my way. I would not have enjoyed my career in education as much as I have without her constantly by my side taking care of me.

INTRODUCTION

There are seven key recommendations that teachers can readily implement in their classrooms to help students organize their learning so they can study more effectively and efficiently, resulting in increased student achievement. The seven—or as I think of them, the "6+1"—connect preparation and instruction to student notes, homework assignments, test preparation, and assessment. The "+1" refers to the importance of building stronger, positive student-teacher relationships.

As you read the book, you can't help but experience déjà vu moments because of the interconnectedness of the recommendations. The very simple idea behind these strategies is to assist students in organizing their learning so they can study effectively and efficiently in a very focused way, so emphasis is placed on what we expect them to know, recognize, and be able to do.

Simply stated, the recommendations I make with respect to preparation will reverberate through the recommendations dealing with instruction, note-taking, homework assignments, test preparation, and tests. The notes, planned in advance will be impacted by the instruction, as will the homework, practice tests, and tests. These recommendations are inseparable.

The beauty behind these seven is that teachers already implement all these recommendations to some degree. The 6+1 refine these accepted practices to help struggling students and students living in poverty feel more comfortable in their knowledge, understanding, and application of mathematics and to answer the question that parents want to know: What are you doing to help my child learn?

The book is organized around the 6+1. Teaching is not as easy as some would have you believe. In fact, teaching is very complex, but implementing the recommendations will provide teachers a needed framework to optimize the learning of struggling students. The recommendations also have the added benefit of keeping students on task and engaged, which diminishes issues in classroom management. And most importantly, implementing these strategies will have an immediate impact on student performance.

At the end of each chapter is a Points to Remember section that reviews the main recommendations in the chapter, the "look-fors" when observing classroom instruction.

Chapter 1 introduces these accepted teacher protocols and expectancies and discusses the significance of a balanced delivery of instruction and assessment. A great deal of emphasis is placed on linking concepts and skills to previously learned math and outside experiences. By introducing new topics through linking, teachers have an opportunity to review, reinforce, and address student deficiencies while making students more comfortable learning the new concepts and skills by using familiar language and experiences.

Chapter 2 stresses the importance of planning and preparation and how they impact almost everything that happens in a classroom and student achievement. Preparation and planning clearly affect instruction, classroom management, and student success. The next chapter then highlights what a balanced delivery of instruction and assessment, components of an effective lesson, and teacher expectancies might look like in a math classroom and how those enhance instruction and student understanding and establish routines that clearly assist struggling students.

Chapters 4 and 5 address support structures that help students organize their learning so they can study more effectively and efficiently. Like the components of an effective lesson, these structures are typically already in place in most math classrooms and just need refinement to help students learn.

Chapter 4 addresses the need for improved student notes that support and reflect instruction and that students can use to complete their homework, study for unit tests and quizzes, and prepare for high-stakes tests such as end-of-year exams. Chapter 5 underscores the critical role more appropriate homework assignments play in monitoring student learning and how the homework supports and reflects student notes and instruction.

Chapters 6 and 7 focus on assessing student knowledge. Strategies are introduced to better prepare students for tests and to connect that test preparation to reflect and support homework, notes, and instruction, linking them all together to support student study and increased student achievement.

Chapter 8 examines the role of positive student–teacher relationships and links those relationships to students' willingness to work for teachers. Chapter 9 then addresses steps that can be taken by teachers and administrators to improve instruction, making students more comfortable in their knowledge, understanding, and application of mathematics and resulting in increased student achievement.

Finally, chapter 10 provides a brief summary of the recommendations that organize student learning by connecting preparation and instruction to student notes, homework assignments, test prep, and tests that support and reflect instruction, as well as the important role relationships play in increasing student achievement.

The recommendations in this book are not designed to be cherry-picked—they cannot be taken in isolation. They are connected to what we do to help students learn. Teachers already employ many of these recommendations to some degree and view them as accepted practices. To truly help students feel more comfortable in mathematics, these educational protocols and expectations, refinements to accepted practices, should become non-negotiable expectations so our struggling students who continually earn grades of D or F begin to experience success because their teachers have organized their instruction to help them not only to succeed but to excel in mathematics.

1

PROTOCOLS AND EXPECTATIONS NEEDED FOR STUDENT SUCCESS

As in many professions, there are protocols and expectations that professionals follow to ensure their clients are receiving the full benefit of their expertise. It is no different in education. There are protocols and expectations to which school administrators and classroom teachers adhere.

When observing math instruction, we expect lessons to be designed to help student learn. Those lessons are made of several components, which will be referred to as the Components of an Effective Lesson (CELs). Those components are typically introduced sequentially and are a part of most math lessons; they are the math teachers' protocols.

The CELs consist of the following:

- Introduction
- Daily review
- Stated daily learning objective
- Concept and skill development and application (instruction)
- Practice (guided/independent/group)
- Homework assignments
- Closure
- Long-term memory review

While these may have different names in different schools, each of these components can be observed in most math classes almost every day. They will be discussed in much greater detail in later chapters.

TEACHER EXPECTANCIES

Along with these protocols, there are a number of undertakings we expect of teachers. For instance, we expect them to teach in ways that lead to increased understanding and to make students more comfortable in their knowledge, understanding, and application of mathematics, resulting in increased student achievement. These practices have come to be known as *teacher expectancies.*

These teacher expectancies, while not exhaustive, are embedded in the CELs and consist of the following:

- Employing the building-of-success-on-success model
- Monitoring student learning
- Developing positive student–teacher relationships
- Requiring student notes
- Using simple, straightforward examples
- Emphasizing the importance of vocabulary and notation
- Stressing reading and writing
- Memorizing facts, formulas, and procedures
- Providing opportunities for practice
- Implementing technology appropriately
- Providing opportunities for problem solving
- Preparing students for tests

BALANCED DELIVERY OF INSTRUCTION

Some states have experienced "math wars" between so-called traditionalists and constructivists on the best way to teach math that will result in increased student achievement. They saw it as an either-or proposition. But research has shown that programs that are very successful have a balance in the delivery of instruction matched by a balance in assessment. "Balance" is defined by what we say we value in math education—what we want students to know, recognize, and be able to do.

Balance in instruction and assessment is critical in the following areas, which will be described further below:

1. Vocabulary and notation
2. Conceptual development and linkage
3. Problem solving
4. Appropriate use of technology
5. Memorization of important facts and procedures

VOCABULARY AND NOTATION

Teachers should be encouraged to introduce new concepts and skills by linking those concepts to previously learned knowledge using language students are more familiar and comfortable with. Having said that, a certain amount of thoroughness, precision, and formality is required in mathematics, specifically in terms of notation and vocabulary; these are the building blocks of concepts, and therefore their correct use is vital. Initially introducing new concepts and skills in familiar language should be encouraged by linking, but by the end of the lesson, more formal language should be used to describe the mathematics.

Mathematics notation is a system of shorthand for the language of mathematics. This notation utilizes symbols to denote quantities, relationships, and operations and has evolved over time to enable us to show the manipulation of data and ideas more readily. Notation enables us to designate mathematical concepts and processes with precision and clarity.

Studying test results and student work would suggest to even the casual observer that students miss far too many questions because they simply do not know what is being asked. For instance, to find a degree of a monomial such as $2x^3y^4z^5$, all students need to do is add the exponents. The answer is 12—not very difficult, mathematically speaking. But many students will miss such an easy question. Why? Because they didn't know or understand the vocabulary.

Vocabulary and notation are important and need to be taught explicitly, used in teaching, and assessed on tests. Just as stressing vocabulary and notation—language acquisition—is recognized as being important when addressing the needs of English-language learners, it is also important for addressing the needs of all students studying mathematics. There is no more single important factor that leads to comprehension than acquiring the necessary vocabulary.

Teachers should take great care in developing good definitions. When first learning to add fractions, many students mistakenly add both the numerators and the denominators together. Teachers normally respond by telling the students to add only the numerators and give no further explanation. In order to develop understanding properly, however, a fraction needs to be defined as part of a whole—a numerator and a denominator. The denominator tells you how many equal parts make one whole unit; the numerator tells you how many equal pieces are under consideration. By explaining the definition, when students then try to add denominators, the teacher can then have them analyze their work based on the defined terms and explain that if they added the denominators they would not have one whole unit.

Knowing and understanding vocabulary and notation require teacher modeling, use, reading, and writing. There is, and should be, an expectation that students can understand, read, speak, and write mathematics. Students in elementary school should be able to read 16.023 as sixteen and twenty-three thousandths—not sixteen point zero, two, three. Similarly, secondary school students should be able to read $_nP_r$ as a permutation of n things being taken r at a time—not as "npr."

It is startling to realize how much information is not transferred formally to students. For instance, many teachers in elementary schools teaching subtraction don't use words like *minuend*, *subtrahend*, and *difference* to describe numbers in that operation. Not only don't they use that vocabulary, they don't explicitly teach their students words or phrases that would generally indicate a subtraction problem. Words such as *difference, left,* and *more*, or words in the comparative form ending in *-er* suggest subtraction. Without repeated exposure to vocabulary and notation, students will not acquire the language of mathematics.

Another simple example to emphasize this point might be to ask students to translate the following expression to mathematics: "four less than a number." Many students are very literal, and since they see the four first, they will write $4 - x$. Others might write it as $4 < x$, which is actually translated as "4 *is* less than x." The correct way to translate "four less than a number" is $x - 4$.

Clearly, these fall under the category of language acquisition. Students not acquiring the vocabulary and notation will have great difficulty learning mathematics, and that difficulty will be reflected on high-stakes tests. And teachers need to remember, this is not just a problem for non-English speakers, it's a problem for all students. While we often hear of the difficulties English-language learners (ELL students) encounter while learning, we seldom hear about the problems all students have learning math as a language. All students need to be identified as "math-language learners."

The research is clear: There is no single more important factor that affects student achievement than the acquisition of vocabulary and notation.

CONCEPT DEVELOPMENT AND LINKAGE

Remember the Law of Cosines? No? Then you illustrate my next point: It is not a matter of *if* students are going to forget much of the information they learn, it's a matter of *when* they will forget it. If you were to ask many math teachers the sine of 30° right now, the fact is that many will not be able to immediately recall it. But, since they understand how it was defined and developed, give them a few seconds and they can tell you the answer. Understanding and being able to reconstruct information is important in learning and maintaining knowledge and skills over time.

In the late 1980s and early 1990s, there were math educators who maintained that any type of drill kills. But others suggested that repetition is the mother of learning. If we want students to be problem solvers and critical thinkers, then they must be able to immediately recall important facts and procedures. Memorization is important, and teaching students how to memorize will help them in their learning, thinking, and problem solving.

Now, let's look at these components of balanced instruction and assessment individually and see how nicely they all really fit together so students have a greater appreciation of mathematics.

Concept Development

In classrooms lacking sufficient concept development, teachers primarily emphasize memorization of rules and algorithms with little or no attempt made to help students understand the "why" of mathematics processes. Without a thorough understanding of the underlying mathematical concepts, students are not truly learning mathematics. Mathematics becomes an arbitrary set of isolated rules that don't make sense and have a tendency to look like magic. Teachers must do more than just give the rules; they must build student competence through understanding.

Think of the rules we give students. Two negatives are a positive—unless you are adding, then they are negative. Any number to the zero power, except zero, is one. When you add integers, sometimes you add, sometimes you subtract. When you subtract integers, you change the sign and add, but then again, you might subtract. You can't divide by zero. When you divide fractions, you flip and multiply. Who's making up these rules? Were they out of their minds? Standing alone, the rules don't make much sense.

The fact is that those rules, formulas, procedures, and theorems are just shortcuts that allow students to solve problems or compute quickly. Some of the shortcuts we are most comfortable and familiar with and use often don't make sense either. For instance, asked to multiply 72 by 10, you'd get 720 very quickly. If asked how you arrived at the answer, many would respond by saying they "added a zero." That's not true. If zero is added to 72, the result is still 72. What really happened that allowed you to instantly perform that computation was a pattern was identified.

As mathematics becomes more abstract, "math anxiety" may develop if shortcuts (rules, theorems, and algorithms) have not been developed with an understanding of why they work. If this is not addressed, students can become frustrated and eventually quit enrolling in mathematics classes, even if the grade they earned in their last class was average or above. Each lesson should build on and strengthen the students' mathematical foundation. Teachers should not assume that students have already seen, let alone remember, an explanation of a particular mathematics concept. Even if they have, a quick refresher may be beneficial.

For example, finding the sum of the interior angles of a triangle might be introduced by having students draw a triangle, cut out the three angles of their triangle, and piece them together to form a straight angle (180°)—suggesting correctly that the sum of the three angles is 180°. The Pythagorean Theorem might be introduced by examining the areas of the squares formed by the sides of the triangle. Seeing these patterns, students might hypothesize the area of the square formed by the hypotenuse is equal to the area of the squares formed by the other two legs. Teachers should not just draw a right triangle, identify the hypotenuse, give the formula, and work out problems.

The understanding gained—in combination with sufficient practice and memorization of important information—gives students confidence in their ability to do mathematics. To deliver this balanced approach, time and thoughtful preparation must be given to each and every lesson. This may require teachers to consult a variety of resources, especially professional colleagues, to find applicable concept development activities.

Explaining the why will often address different learning modalities. For example, having the students cut out the angles of a triangle to see how the relationship came about would address kinesthetic learners and help them remember the theorem. Concept development is important because my belief is it is not a matter of whether students will forget, but rather when. For English-language learners, conceptual development of ideas comes under the heading of "building background." It's a necessary step in language acquisition and understanding for all students.

Procedural knowledge, while important, is not enough. Understanding is important. The following questions came from high school exit exams. Each question measures students' abilities to find the measures of central tendency—often referred to as averages. If student understanding of the mean was only memorized information, students would have difficulty answering some of these questions.

1. Find the mean of the following data: 78, 74, 81, 83, and 82.
2. In Ted's class of forty students, the average on the math exam was 80. Andrew's class of thirty students had an average of 90. What was the mean of the two classes combined?
3. Ted's bowling scores last week were 85, 89, and 101. What score would he have to make on his next game to have a mean of 105?
4. One student was absent on the day of a test. The class average for 24 students was 75 percent. After the other student took the test, the mean increased to 76 percent. What score did the last student get on the test?
5. Use the following graph to find the mean.

The first question tests students' procedural knowledge. The p-score on that particular question was approximately 0.8. That means about 80 percent

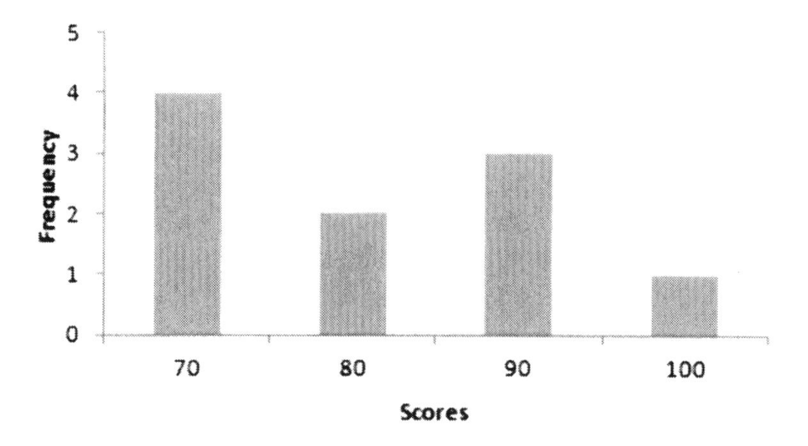

of the students taking the test got the correct answer. They knew procedurally how to find the mean.

On the second question, one of the distracters (possible answers) was 85 percent. The students that relied on only procedural knowledge added 80 to 90, divided by two, and got the incorrect answer of 85 percent. The p-score for that question on the exit exam was 0.48. This is a dramatic decrease from finding the mean in the first question. The question was not very difficult, but apparently students needed greater understanding to answer the question correctly. Students not understanding they had to redistribute the total number of points among the seventy students got that question wrong.

Teachers are asked all the time to determine what grade a student has to make on their next test to earn a particular letter grade. Question 3 is similar. As can be seen from questions 2 and 3, if students do not understand the concept, any variation in problem will cause them difficulty.

Question 4 caused a tremendous amount of difficulty for students. Many incorrectly think the missing student would have to earn a 77 percent on the test rather than 100 percent. Again, not understanding the importance of redistributing the total number of points led these students to the wrong conclusion. Procedural knowledge is important—but so is understanding. The p-score on this question dropped to 0.22. What we can quickly see is that, without understanding, the slightest variation in a problem will cause students a great deal of trouble.

Students often interrupt their teachers as they are trying to develop a concept. Why do they interrupt? What's the big deal? The answer is too easy.

Typically, they want to know tonight's homework assignment. Students know what their teachers value, even more so than the teachers themselves. Students know that if you are not going to ask them a conceptual understanding question on their homework, quiz, or on the test, it must not be important.

Since students value what teachers grade, concept development and linkage should also be tested. Students should be asked to write a brief explanation of a particular concept as part of the homework assignment and then be given an open-ended question on a test where they must explain the origin of a rule or algorithm. If students are tested on the why of mathematics, they will be less likely to tune teachers out during concept development. Balanced delivery of instruction requires balanced assessment.

Linkage

As teachers teach mathematics, they should remain cognizant of the fact that the concepts and skills they are teaching will be used later as building blocks to introduce more abstract concepts. Middle school teachers use concepts and algorithms taught in elementary school, and high school teachers continue to build on student knowledge gained in middle school. This process is referred to as "linkage" or "connections," the introduction of new material through the use of skills and concepts that have previously been taught. The idea of linkage can also be applied to smaller units of time, including material learned yesterday, last week, or last month.

Therefore, as new lessons are prepared and presented, teachers should build links to the new material from previously learned concepts or outside experiences. By introducing concepts through the utilization of linkages, teachers enable students to place new ideas into a context of past learning.

Students introduced to new or more abstract concepts using familiar language are not as threatened. Teachers, on the other hand, have an opportunity to review and reinforce previously learned topics—topics and skills they often identify as deficiencies and reasons why they are not successful teaching their assigned curriculum. Teachers can then compare and contrast that information, and students see the idea in a different context. Simply put, students are then more likely to understand and therefore absorb new material when linkage is being used.

The importance of linking concepts and skills to previously learned material and outside experiences cannot be overstated. Many of our best students probably don't know how the slope formula, and the equations of lines—point-slope, slope-intercept, and general form of the equation of a line—are related.

By not introducing these concepts through linking, teachers lose valuable instructional time. They also lose opportunities to introduce the new material using language students are most familiar and comfortable with and to address deficiencies by reviewing and reinforcing previously taught material. Students are denied opportunities to increase their understanding by comparing and contrasting those ideas, as well as not seeing the math used in different contexts. Most teachers cannot do this on the fly. Preparation matters.

As another example, rather than just having students "flip and multiply" when dividing fractions, the division algorithm might be developed through repeated subtraction—just as was done in fourth grade with division of whole numbers. Solving equations should be connected to the Order of Operations. The standard multiplication algorithm that is taught in fourth grade is the same algorithm that is used in algebra to multiply polynomials. Invariably, student memory, over time, will diminish. An understanding of where theorems, formulas, and algorithms (shortcuts) originated will enable students to reconstruct concepts and solve problems.

Where possible, linkages should also be made between concepts within the course as well as to student experiences in "real life." Buying candy at a store can be linked to such mathematical concepts as ratios, proportions, slope, ordered pairs, graphing, and functions. Students quickly see that if one candy bar costs fifty cents, then two will cost a dollar. The connection is readily translated to the math they learn in the classroom. As a proportion, 1 candy bar is to $0.50 as 2 candy bars is to $1.00; or, written as ordered pairs: (1, .50), (2, 1.00). Linking makes math more relevant—and is very important for students trying to learn the language or for students of poverty—by reviewing and reinforcing previously learned concepts and skills in a nonthreatening manner.

Buying dirt for a garden or constructing a patio can be linked to volumes of rectangular prisms. The circumference of a circle can be linked to pipe fitting or the odometer and speedometer readings on a car. Systems of equations can be linked to decision making. The links to make math come alive are too numerous to mention. Teachers must transfer their knowledge, interest, and enthusiasm in mathematics to their students. That can't be done by doing rote problems, without seeing the benefit of math in solving everyday problems that too often are not even considered mathematics.

Too many algebra students are asked to do the mundane. When working with conics, they might be asked to find things like the vertex, directrix, and focus of a parabola, and that's great. But what would be even better is if, while they were studying the parabola, the teacher made students aware of how they use the concept of parabolas in their everyday life. Flashlights, headlights on a car, a satellite dish, and amphitheaters are all examples of how that mathematical concept is used.

More Examples of Linking

Introducing new concepts and skills through linking is so important to increasing student achievement, especially for students coming from poverty. Here are a number of examples that can be immediately employed in the classroom.

Besides linking new material to previously learned material, it is helpful to link it to outside experiences as well. The idea of *slope* is used quite often in our lives, for instance, although outside of school it goes by different names. People involved in home construction might talk about the pitch of a roof. Riding in a car, students might have seen a sign on the road indicating a grade of 6 percent up or down a hill. Both of those cases refer to what we call slope in mathematics.

Kids use slope on a regular basis without realizing it. Returning to the candy bar example, a student buys a candy bar for fifty cents; if two candy bars were purchased, the student would have to pay a dollar. That could be described mathematically by using ordered pairs: (1, $0.50), (2, $1.00), (3, $1.50), and so on. The first element in the ordered pair represents the number of candy bars; the second number represents the cost of those candy bars. Easy enough, don't you think?

Now, if the students were asked how much more was charged for each additional candy bar, we hope the students would answer fifty cents. The difference in cost from one candy bar to another is $0.50. The cost would change by \$0.50 for each additional candy bar. Another way to say that is the *rate of change* is \$0.50. In math, we call the rate of change "slope."

The slope is often described by the ratio $\frac{rise}{run}$. The "rise" represents the change (difference) in the vertical values (the *y*'s); the "run" represents the change in the horizontal values (the *x*'s). Mathematically, we write

$$m = \frac{y_2 - y_1}{x_2 - x_1}$$

Let's look at two of those ordered pairs from buying candy bars—(1,\$0.50) and (3,\$1.50)—and find the slope. Substituting in the formula, we have:

$$m = \frac{\$1.50 - 0.50}{3 - 1} = \frac{\$1.00}{2}$$

Simplifying, we find the slope is \$0.50. The rate of change per candy bar is \$0.50.

The importance of using linkages to introduce new or more abstract concepts and skills cannot be overstated. Linking allows teachers to introduce new ideas in familiar language in a nonthreatening manner, to review and reinforce that knowledge, to compare and contrast, and to see mathematics used in different contexts.

PROBLEM SOLVING

Mathematics is more than just memorizing rules and procedures. It is a discipline, a way of thinking. Students must be taught and encouraged to think, to imagine, and to be creative in their approaches to solving problems. As the National Council of Teachers of Mathematics (NCTM) states, "Problem solving is not a mystery." It is also not limited to solving traditional story problems or word problems. It is a way of thinking that can be learned.

Teachers need to encourage their students to approach learning and problem-solving activities with an open mind and to realize that this kind of thinking takes time and effort to achieve. Students' answers, whether correct or not, should be viewed as opportunities to explore thinking strategies. Open-ended questions that evoke thoughtful responses and require more than one-word answers should be presented. Students should also be encouraged to utilize a variety of problem-solving methods. While problem solving is difficult to teach, requiring commitment and patience on the part of both teacher and learner, it is an essential experience.

Requiring teachers to have a balance in their delivery of instruction and assessment should come under the heading of teacher expectancies. In math, we expect all teachers to employ such recognized problem-solving and learning strategies as:

- Go back to the definition
- Look for a pattern
- Make a table or list
- Draw a picture
- Guess and check
- Examine a simpler case
- Examine a related problem
- Identify a subgoal
- Write an equation
- Work backward

These problem-solving strategies also help students understand the mathematical concepts being taught. It's no mistake that "Go back to the definition" is listed first. Too often it is not listed as a problem-solving strategy. But without a good definition, without knowledge of vocabulary, students are bound to encounter difficulty in any subject. It is difficult to understand how teachers can explain to students not to add the denominators when adding fractions without a good definition of a fraction.

Successful teachers are cognizant of the problem-solving and learning strategies in their daily instruction. When students encounter difficulty in understanding a concept or mastering a skill, good teachers encourage students to go back to the definition, draw pictures, look for a pattern, or examine a simpler or related problem. Students who have been taught and encouraged to use such strategies don't sit idly when they encounter difficulty; they are better prepared to address the problem at hand.

Some may wonder why some students are successful in one subject but not in another. For example, we often see students successful in an algebra class having great difficulty in geometry. Why is that?

Algebra teachers tend to use the strategies of *look for a pattern*, *make a table*, *examine a simpler case*, and *write an equation* as the basis for most of their instruction. As a result, students grow comfortable learning math with these strategies. Unfortunately, too many students still try to learn algebra by rote memorization, so any variation of a problem causes great difficulty and frustration for them. That can be clearly seen on exit exams in mathematics.

Teachers of geometry tend to use *go back to the definition*, *draw a picture*, *examine a related problem*, *identify a subgoal*, and *work backward* as their primary strategies. Students who learned algebra by memorizing often run into difficulty in geometry. Students and teachers who use the same strategies to teach or learn geometry that were successfully used in algebra often run into difficulty, too—resulting in higher fail rates.

In geometry, students are typically required to use higher-order thinking skills that are not being used in a typical algebra class. Experience tells us that too many geometry students do not have a good visualization of the definitions, postulates, and theorems that are being introduced. As we expect students to learn these, they should also be able to draw a picture that reflects the information being taught as well as to measure the drawings and diagrams to explore, discover, and eventually commit to memory important theorems and procedures.

Students should be required to write their definitions, postulates, and theorems on their homework, quizzes, or tests. They should also be required to draw a corresponding picture. If they do that and visualize the information, they will be more successful learning geometry.

Geometry is filled with new terminology and notations, and teachers need to be mindful that student success in any subject is dependent on them learning the language. All too often in math, the difficulties experienced by students have more to do with a lack of understanding of vocabulary and notation than of the math concept being taught. Classroom teachers should take the time to ensure students are learning and using that vocabulary and notation, and they should also be testing students on it.

And while some problem-solving and learning strategies are used more routinely in one subject area than another, the fact is that all of these strategies should be used at appropriate times in all of mathematics.

And finally, let's not forget about "linking" geometry to algebra—referred to as coordinate geometry. Linking allows teachers to introduce new concepts in familiar language, to review and reinforce, to compare and contrast, and to teach in a different context—all of which, the research suggests, leads to increased student achievement. While coordinate geometry is typically a chapter by itself toward the end of a book in geometry, these links can and should be made all year.

APPROPRIATE USE OF TECHNOLOGY

The NCTM observes that "the thoughtful and creative use of technology can greatly improve both the quality of the curriculum and the quality of children's learning. Integrating calculators and computers into school mathematics programs is critical in meeting the goals of a redefined curriculum." However, the NCTM also says, "Calculators do not replace the need to learn basic facts, to compute mentally, or to do reasonable paper-and-pencil computation." Therefore, appropriate use of technology is dependent on the age of a student and his or her ability to demonstrate knowledge of basic facts. It is further dependent on the objective of the activity. If the goal is skill attainment, then calculator use is not appropriate. But if the goal is exploration or verification, then calculator use may be appropriate.

Modern technology can free students from tedious computations and allow them to concentrate on problem solving and other important mathematics content. Students should be using calculators to strengthen and extend understanding of concepts, explore mathematical functions, investigate problem-solving activities, employ real-world applications, and verify results. In Algebra I and above, the use of graphing calculators is imperative. However, it is equally essential that all teachers maintain a balance between paper-and-pencil computation or drill and the use of technology to enhance problem solving and conceptual learning.

This requires teachers to make a conscious decision as to the appropriateness of calculator use during each and every lesson. Calculators should not be allowed as a substitute for thinking. To increase the likelihood that calculators will be used appropriately, teachers may need additional training. Total dependence on technology is inappropriate, but when combined with an understanding of the underlying concepts and proficiency with basic skills, technology becomes an invaluable tool.

For example, one method to find the zeros of a quadratic function is through factoring. Students should easily be able to find the zeros of the function $f(x) = x^2 - 4x - 12$ without a graphing calculator, as follows:

$$x^2 - 4x - 12 = 0$$
$$(x - 6)(x + 2) = 0$$
$$x - 6 = 0 \text{ or } x + 2 = 0$$
$$x = 6 \text{ or } x = -2$$

Students should have learned, however, that the graph of the quadratic function $y = x^2 - 4x - 12$ has x-intercepts at $x = 6$ and $x = -2$, indicating the function's zeros. A quick method for finding the location maximum/minimum of the parabola defined by $y = x^2 - 4x - 12$ is to average the zeros In this example, the minimum occurs at $x = \frac{6+(-2)}{2} = 2$. The y-coordinate of the extremum is $f(2) = 2^2 - 4(z) - 12 = -16$. Again, students should have the ability to do this without the graphing calculator.

A graphing calculator is an ideal tool for exploration of the following problem:

A diver stands on the 7-meter diving platform preparing for a dive (the 7-meter platform is 7 meters above the surface of the water). The diver jumps vertically upward from the platform with an initial velocity of 5 meters per second. The diver's height above the water can be modeled by the equation $h = 7 + 5t - 4.9t^2$, where t is the elapsed time in seconds since the diver jumped. At what time after the dive does the diver reach the water? At what time does the diver reach a maximum height and how high is it?

The expression $7 + 5t - 4.9t^2$ is not easily factored, nor is completing the square trivial. The quadratic formula leads to solutions, but a graphical exploration yields more information and eliminates tedious calculations (see figure 1.2). Students should determine that the diver reaches maximum height of 8.28 meters at $t = 0.51$ second and enters the water at 1.81 seconds. Further exploration may prompt discussion about how high the diver jumped above the platform and the meaning of the quadratic's other zero at $t = -0.79$ second.

To think critically and to problem-solve, students need understanding and a body of information to draw from. To do those, they need vocabulary and to be able to use technology. Balance in the delivery of instruction and in assessment ensures that our students are getting a full, rich curriculum. It also ensures that students are being taught what the adults say they value in mathematics.

MEMORIZATION OF IMPORTANT FACTS AND PROCEDURES

Mastery of basic facts is an essential part of learning mathematics. When students encounter mathematics concepts, they need instant recall of basic facts. Stopping to remember these facts interrupts the flow of thought, which negatively impacts learning.

What constitutes "basic knowledge" depends on the grade level. Basic facts in elementary school might be arithmetic facts. In middle school, they might be expanded to include the conversions between fractions, decimals, and percentages or the algorithm for adding fractions. In high school, basic facts may also include the quadratic formula, the Pythagorean Theorem, knowing what the graph of a second-degree polynomial equation looks like, or algorithms for solving linear equations.

Since student deficiencies are evident at all levels, teachers should regularly revisit basic facts. Many higher-level thinking processes required for success in high school mathematics courses demand immediate recall of basic facts. The demands of teaching dense curricula and addressing student deficiencies may, at times, overwhelm the teacher. However, if carefully analyzed and incorporated into lesson plans, deficiencies can be addressed successfully.

The most common complaint heard from teachers is that they can't teach their curriculum because their students don't know their basic math facts. Sixth-, seventh-, and eighth-grade teachers will then spend the first three or four weeks of the school year reviewing and reteaching arithmetic facts.

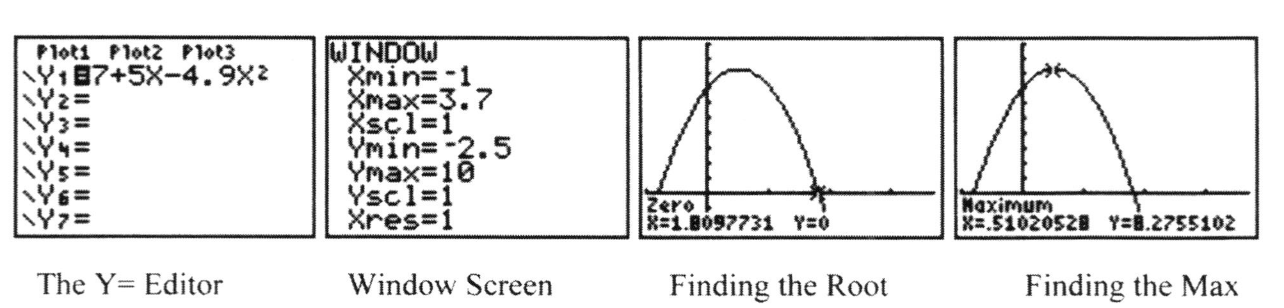

Figure 1.2. A quadratic equation solution on a graphing calculator

Those teachers continually complain about how the elementary teachers did not do their job. If you do the math, you can quickly see that approximately nine weeks of middle school is spent addressing a nonexistent problem. Nonexistent?

Many teachers, well-meaning as they are, review all 100 multiplication facts. But if they based their instruction on data, most of those teachers would have found out most kids knew their ones, twos, threes, fives, tens, doubles, and nines. By incorporating the commutative property, teachers would find there are only about seventeen combinations in which the kids are really experiencing difficulty. Rather than pulling out all the flash cards, students might be better served if their teachers concentrated on their deficiencies and reviewed the other facts.

If students are struggling with their basic addition and multiplication facts, the following strategies for teaching basic arithmetic facts may be of help. What's important to note in these strategies is that the facts are not necessarily taught sequentially. These arithmetic facts, like anything else we teach, should be taught in a manner and order that helps students learn.

Additionally, students in elementary schools should be spending seven to twelve minutes almost daily reviewing their basic arithmetic facts. If that was a common practice and teachers modeled those strategies regularly, the problems with deficiency could be minimized.

Secondary teachers were not trained, and most don't know the strategies, for teaching basic arithmetic facts. They typically tell the students to go home and memorize them. That does not result in much success. The following strategies should assist all teachers to help their students learn the basic arithmetic facts with fluency—automaticity.

Strategies for Learning Addition Facts

There are 100 basic addition facts with sums to 18 (see figure 1.3). By using commutativity ($3 + 7 = 7 + 3$), we can reduce the total number of addition facts students must learn to fifty-five.

1. *Adding zero:* Students quickly understand that the sum of zero and any number is that number. For example, $0 + 6 = 6$. (This leaves forty-five addition facts students must memorize.)

+	0	1	2	3	4	5	6	7	8	9
0	0	1	2	3	4	5	6	7	8	9
1	1	2	3	4	5	6	7	8	9	10
2	2	3	4	5	6	7	8	9	10	11
3	3	4	5	6	7	8	9	10	11	12
4	4	5	6	7	8	9	10	11	12	13
5	5	6	7	8	9	10	11	12	13	14
6	6	7	8	9	10	11	12	13	14	15
7	7	8	9	10	11	12	13	14	15	16
8	8	9	10	11	12	13	14	15	16	17
9	9	10	11	12	13	14	15	16	17	18

Figure 1.3. Basic addition facts

2. *Counting on by 1 and 2:* Students often find sums with addends of 1 or 2 by simply "counting on." (This leaves only twenty-eight facts left to learn.)
3. *Sums to 10:* Students can readily identify sums to 10 by repeated experiences with 10. (This leaves twenty-five facts.)
4. *Doubles:* For whatever reason, students seem able to remember doubles—for example, 7 + 7 = 14—more easily than other combinations of numbers. (Now, nineteen facts are left.)
5. *Doubles plus 1:* Adding consecutive numbers. Knowing that 7 + 8 is equivalent to 7 + (7 + 1) helps students remember these sums. (This leaves thirteen facts.)
6. *Doubles plus 2:* Adding consecutive odd or even numbers. Knowing that 5 + 7 is equivalent to 5 + (5 + 2)
7. *Adding nines:* Students can quickly see that when they are adding the units digit in the sum is one less than the number they are adding to 9. For example, 7 + 9 = 1*6* since the 6 in the units place is one less than 7. (Only eight facts are left.)
8. *Adding tens*

Thinking Strategies for Learning the Subtraction Facts

1. *Fact families:* This strategy is the most commonly used and works when students understand the relationship between addition and subtraction, that is, when students see 6 – 2 and think 2 + ? = 6. However,

if this strategy is used with the following strategies, students will find greater success in a shorter period of time.

2. *Counting backwards:* This method is similar to "counting on" used in addition, but it isn't quite as easy. Some might think if you can count forward, then you can automatically count backward. This is not true—try saying the alphabet backward. Students should only be allowed to count back at most three numbers.
3. *Zeros:* The pattern for subtracting zero is readily recognizable: 5 – 0 = 5.
4. *Sames:* This method is used when a number is subtracted from itself; this is another generalization that students can quickly identify: 7 – 7 = 0.
5. *Recognizing doubles:* Recognizing the fact families associated with adding doubles.
6. *Subtracting tens:* This is a pattern that students can pick up on very quickly, seeing that the units digit remains the same.
7. *Subtracting from ten:* Recognizing the fact families for sums to 10.
8. *Subtracting nines:* Again, the pattern that develops for subtracting 9 can be easily identified by most students. They can quickly subtract 9 from a minuend by adding 1 to the units digit in the minuend: 17 – 9 = 8, 16 – 9 = 7.
9. *Subtracting numbers with consecutive units digits:* This pattern will always result in a difference of 9, 16 – 7 = 9, 13 – 4 = 9, 15 – 6 = 9—all have units digits that are consecutive and the result is always 9.
10. *Subtracting numbers with consecutive even or consecutive odd units digits:* This pattern will always result in a difference of 8: 14 – 6 = 8, 13 – 5 = 8, 12 – 4 = 8.

These strategies clearly help students to subtract quickly. Allowing the students see the patterns develop will make students more comfortable using these shortcuts and get them off their fingers.

Having said that, without being able to identify the proper strategy by examining the problem, memorizing these strategies may become more burdensome and cause greater confusion than just rote memorization. As with many of the concepts and skills in math, students need to compare and contrast problems to make them more recognizable to them.

So while you might teach one strategy at a time, as you add to the number of strategies students can use for a specific numbers, you will need to review previous strategies and—this is important—combine strategies on the same work sheets, asking students only to identify the strategy they would use for each problem and why they would use it. Being able to compare and contrast will lead to increased student understanding, comfort, and achievement using these strategies. For example:

- 16 – 9: Students are subtracting 9, so they add one to the units digit.
- 15 – 7: Students are subtracting numbers with consecutive odd units digits, so the difference is 8.
- 17 – 8: Students are subtracting numbers with consecutive units digits, so the difference is 9.

Trying to learn basic math facts by just memorization, rereading the tables, or using flash cards is a painful experience for most students and is just as agonizing to watch. Memorization and using flash cards are necessary, but how teachers teach those facts can have a significant impact on students' success.

For instance, many teachers use the "families" for students to memorize the subtraction facts. That is, if 5 + 4 = 9, then 9 – 4 = 5. That's all right, and those relationships should be developed. However, students should be taught when they subtract 10, the units digit remains the same in the difference. They should also be taught to recognize that when subtracting 9, the units digit is always one more than the units digit in the minuend.

There are other patterns that might be developed that would help students learn their subtraction facts. For example, when you subtract consecutive numbers—that is, when the units digit in the subtrahend is one larger than the units digit in the minuend—the answer is always 9. For example, 16 – 7 = 9 and 14 – 5 = 9.

Continuing this hunt for patterns, when you subtract consecutive odd or consecutive even numbers, for example, 13 – 5 or 14 – 6, the difference is always 8. Helping students recognize those patterns will accelerate their learning of the math facts and will relieve some of the pain often felt by these students.

Algorithms

Procedural fluency has been identified in the research as being important to increase student achievement in mathematics. The use of algorithms—systematic, step-by-step procedures used in computation or problem solving—helps to address the difficulty students often have sequencing complex mathematics problems. The NCTM's *Curriculum and Evaluation Standards* recommends that students use algorithms to compute and solve problems. However, algorithms should not stand alone and usually need to be preceded by concept development.

By developing an understanding of a concept, students will be better able to understand the objective involved. They will then be more willing and able to identify patterns that lead to the shortcuts we call rules, algorithms, formulas, theorems, or conjectures. These shortcuts were developed in many instances because someone recognized a pattern that would give them the desired result without having to do as much work. Teachers should stress to students that the shortcuts, by themselves, often do not make sense. It is vital that students understand the concepts and how and why the shortcuts work.

While there are many different ways to compute, and therefore many different algorithms, the teaching of standard algorithms is important because this ensures that students have common frames of reference. This is significant, as the standard algorithms developed in elementary grades become the foundation upon which more abstract material is introduced in middle school.

For instance, there are many ways of multiplying. The ancient Egyptians used an algorithm known as repeated doubling; fourteenth-century Italians used the "lattice method." In the United States, to maintain consistency, we have identified a standard multiplication algorithm. A variation of this algorithm is later used in algebra when multiplying two binomials (the FOIL method).

In another example, the standard division algorithm taught in fourth and fifth grades is used again in algebra when students divide polynomials. It is also used in synthetic division and synthetic substitution when solving higher-degree equations using the Rational Root Theorem. The division algorithm is important. Teachers expect students coming into their classes will have had certain learning experiences. If students lack practice with the standard multiplication and division algorithms or other standard U.S. algorithms, they will probably experience unnecessary difficulty in future mathematics classes.

Here are a few examples to demonstrate how standard algorithms are continually used in mathematics.

The standard algorithm for multiplying two two-digit numbers is:

1. Multiply the multiplicand by the digit in the units column of the multiplier.
2. Indent from the right a space to account for place value and multiply the multiplicand by the multiplier's tens digit.
3. Add those partial products.

An illustration from fourth grade math would be:

$$\begin{array}{r} 32 \\ \times\ 21 \\ \hline 32 \\ 64 \\ \hline 672 \end{array}$$

A similar process is used in algebra:

$$\begin{array}{r} (x + 4) \\ \times\ (x + 5) \\ \hline 5x + 20 \\ x^2 + 4x \\ \hline x^2 + 9x + 20 \end{array}$$

When young students are asked to model or explain how to determine the number of twos in eight, they might use the repeated subtraction model shown below:

$$8 - 2 = 6,\ 6 - 2 = 4,\ 4 - 2 = 2, \text{ and } 2 - 2 = 0$$

Students can clearly see they subtracted 2 a total of four times. Therefore, there are four 2's in 8.

This same concept applies to division of fractions. If asked to divide $\frac{3}{4}$ by $\frac{1}{8}$, students should be able to use the same repeated subtraction model.

$$\frac{3}{4}-\frac{1}{8}=\frac{5}{8},\quad \frac{5}{8}-\frac{1}{8}=\frac{4}{8},\quad \frac{4}{8}-\frac{1}{8}=\frac{3}{8}$$

$$\frac{3}{8}-\frac{1}{8}=\frac{2}{8},\quad \frac{2}{8}-\frac{1}{8}=\frac{1}{8},\quad \frac{1}{8}-\frac{1}{8}=0$$

Students can see they subtracted $\frac{1}{8}$ a total of six times. Thus, there are six $\frac{1}{8}$'s in $\frac{3}{4}$.

Given opportunities to look for a pattern, for practice and appropriate guidance, students might notice that rather than performing all those repeated subtractions, if they multiplied by the reciprocal of the divisor they would arrive at the desired result. Not only would it give them the correct answer, they would be able to do the computation faster and more efficiently.

That pattern would lead to the following algorithm for the division of fractions:

1. Make sure you have proper or improper fractions.
2. Invert the divisor.
3. Divide out common factors.
4. Multiply numerators.
5. Multiply denominators.
6. Simplify.

As an example:

$$\begin{aligned}\frac{3}{4}\div\frac{1}{8}&=\frac{3}{4}\times\frac{8}{1}\\&=\frac{24}{4}\\&=6\end{aligned}$$

When students discover the patterns derived by playing with numbers through teacher guidance—the process known as "directed discovery"—they can be shown that algorithms are nothing more than a faster way to compute or solve problems by applying those patterns. Mathematics then is no longer something magical or mysterious; it becomes a powerful tool to be used in a variety of situations.

By learning an algorithm, students will have a method with which to solve a variety of problems. Likewise, students that know how to solve a problem should be able to verbalize what they have done, verify and defend their solutions, and communicate results. The memorization and utilization of algorithms allows students to do just that.

Memorization—Oral Recitation

In every course, there are certain items that all students should know at the completion of the school year, information upon which understanding and critical thought can be based. Furthermore, the more sophisticated mental operations of analysis, synthesis, and evaluation are impossible without rapid and accurate recall of bodies of specific information. When these items are first introduced, oral recitation should be utilized to help students memorize the information.

For example, after the initial development of the quadratic formula, an algebra teacher could provide it visually and then orally read the formula several times so the students know how to correctly say it. Following this, the class as a whole could recite it for ninety seconds—an eternity. Finally, individual students could be called upon to recite the formula without the use of a visual aid. This process places the information in students' short-term memory with the expectation that it will then, through repeated and extended exposure, be transferred to long-term memory. Oral recitation may need to be used again when important topics are revisited.

Oral recitation is the practice of having the entire class recite important facts, identifications, definitions, and procedures within the instruction and later when they need to be revisited. Concept development generally precedes oral recitation. Whole-class recitation (repetition) of this information should be repeated a number of times; however, the total time involved should not exceed two and a half minutes. By having the students first read the information off the board with the teacher, students learn how to read the information correctly and how to say it. Oral recitation is a language acquisition strategy that helps all students learn—not just English-language learners.

The use of oral recitation should not lead teachers into the "gotcha" practice. While oral repetition will help most students imbed this information into short-term memory, other students learn differently. Incorporating the idea of building success on success might suggest you don't want to call on students that you think might not have it in short-term memory.

Use oral recitation from a positive perspective. That is, after the class orally recites for approximately two minutes, call on students that you believe can give the information back to you. Those correct responses create the impression with students that other members of the class are getting the information, as opposed to the idea that "nobody is getting it."

Oral recitation is just one method of helping students memorize information. Adults often use it when trying to remember a license plate number or grocery list. This practice anchors information in the brain and helps students absorb and retain information upon which understanding and critical thought is based. Again, the more sophisticated mental operations of analysis, synthesis, and evaluation are impossible without rapid and accurate recall of bodies of specific information. Memorization, while not fun, is important. The process also keeps students engaged in learning, helps them verbalize their knowledge, and suggests that if the information being presented is important enough for the entire class to recite, it is worth remembering.

Memory Aids

> Mnemonics are based on the principle that the brain is a pattern-seeking device, always looking for associations between the information it is receiving and what is already stored. If the brain can find no link or association, it is highly unlikely that the information will be stored in long-term memory. Unfortunately, this scenario is relatively commonplace in the classroom. (Patricia Wolfe, *Brain Matters: Translating Research into Classroom Practice [Alexandria, VA: ASCD*, 2001])

That's another reason linking should be used more often in the classroom.

The brain has trouble storing information that it cannot associate to a picture such as letters and numbers. Mnemonics create rhyming links or associations that give the brain an organizational framework on which to hook new information.

In certain circumstances, teachers use mnemonics because they were taught them while they were in school. Many middle school students are introduced to the phrase "Please Excuse My Dear Aunt Sally" as a way of remembering the Order of Operations; similarly, "SohCahToa" is familiar to many high school students learning the trigonometric ratios.

Helping them remember over time is important if we are to build credibility in the public education system. As students are first being introduced to definitions, concepts, and skills, their likeness causes them confusion. For instance, if we look at the definitions of complementary angles and supplementary angles, the definitions are similar and students might mix them up. A teacher might suggest students associate the "c" in complementary with a corner because it also begins with a "c," hence, 90°. Likewise, the "s" in supplementary might be associated with the "s" in a straight, thereby associating that with 180°.

Memory aids help students study more effectively and efficiently. Used in the classroom, they will help students remember, thereby increasing student achievement.

Building Memories

Instruction matters. How students are taught makes a tremendous difference in students' understanding and comfort in learning mathematics. Organizing student learning on what has been presented by connecting that preparation and instruction to student notes, homework assignments, test preparation, assessments, and student-teacher relationships focuses students on what they need to know. But let's make no mistake— memory is key to learning.

Memorizing can help students absorb and retain information on which understanding and critical thought are based. The more sophisticated mental operations of analysis, synthesis, and evaluation are impossible without rapid and accurate recall of bodies of specific knowledge. Classroom teachers can enhance their students' memories by employing instructional strategies that create, stimulate, and enhance neural pathways in the brain so students can more readily access that information.

By beginning a math class with a quick, crisp, focused, and purposeful review of recently taught material, teachers can further cement into memory what was taught previously. If the teachers were to write on the board a definition, formula, or procedure the students were introduced to the day before, then have the class recite that information for thirty seconds, students would be not only blending learning modalities but also reinforcing their short-term memory. In this review, to expand instructional time, teachers might also include one or two exercises from the previous night's homework to address questions or concerns students may have encountered in completing the assignment.

Brain research suggests the importance of the beginning and ending of an activity with respect to remembering. As an example, if I were to give you a list of nine items and ask you to write them down after orally presenting them to you, chances are you would remember the first couple and the last few items listed—the beginning and the end. Speechwriters have always been conscious of this, making sure a speech has a strong beginning and a strong end. Likewise, teachers should have not only a strong beginning to a class but also a strong closing—a closing that repeats what they expect their students to know, recognize, and be able to do.

When teachers introduce new concepts and skills by linking them to previously learned material or outside experiences, students can use those connections to better define the neural pathways in their brain and help place the information into long-term memory by using contextual memory. For instance, if my wife sends me to the grocery store to buy seven items and when I get there I can remember only five of those items, I have a problem. However, if I knew that the items I was sent to the store to buy were for a turkey dinner, then I could use that contextual memory to help me remember the two items that I forgot.

So, while teaching students to memorize an algorithm for adding decimals may be important, it would be more helpful to students if that algorithm were linked to something else they knew—like adding fractions. That linkage allows teachers to review and reinforce material and fortifies the memory by having a bigger picture of related information rather than a bunch of small snippets.

Teachers who require students to take notes also enhance memory by building patterns or developing concepts using pictures that lead to rules, formulas, and procedures. The visual memory of what they have written in their notes also reinforces their memory.

After new concepts and skills have been developed, teachers should write that information on the board and have the entire class recite the formula, rule, or procedure, first by reading it to them, then having the class recite that information for a minute and a half to two minutes—an eternity. We have had experience with immediate memory and know how quickly it fades. How many times have we called information for a phone number, hung up, and quickly dialed the number before we forgot it?

We need to take new information and strengthen the neural pathways so students can access that information quickly. By writing the developed algorithm on the board with an example problem beside it, students are able to recite the procedure and visualize the corresponding problem at the same time. That double-encodes that information in the brain.

For example, if we develop the algorithm for solving quadratic equations by the zero product property (factoring), we should write the algorithm on the board and place an example right next to it, as shown in figure 1.4.

Typically, after a rule or procedure is developed and placed into memory by recitation, background information, and connections, teachers provide students an opportunity for guided practice. Guided practice and homework assignments further cement that knowledge into memory. Students learn the rule or procedures and then use that information to do problems.

Teachers can implement in their everyday instruction techniques that strengthen and enhance students' capacity to remember information. A practice test also provides a clear blueprint for the students on the "things" they have to know to be successful. For students who have experienced a great deal of difficulty in math, the practice test can act as a motivational tool, remove the unknown, connect what they are learning to what will be tested, and create the sense that if they answer the problems

Algorithm—Solving Quadratic Equations

Find the solution set for $x^2 = x + 20$
Use the Zero Product Property

1.	Given	$x^2 = x + 20$
2.	Place everything on one side, zero on other side	$x^2 - x - 20 = 0$
3.	Factor completely	$(x - 5)(x + 4) = 0$
4.	Set each factor equal to zero	$x - 5 = 0, x + 4 = 0$
5.	Solve the resulting equations	$x = 5, x = -4$
6.	Write as a solution set	$\{5, -4\}$

Figure 1.4. Sample algorithm

on the practice tests, they can be successful—a light at the end of the tunnel instead of despair.

POINTS TO REMEMBER

Not only is *what* we teach important, but *how* we teach it also affects student learning. We need to look at proven instructional strategies that help students learn.

To teach math successfully, students must experience a full, rich, and more in-depth curriculum—a balanced delivery of instruction. That balanced delivery must be accompanied by assessments that reflect that balance.

There's a lot to teaching. It is more than just going into a classroom and talking. It requires forethought and planning to ensure students are provided opportunities not only to learn math but also to feel comfortable in their knowledge, understanding, and application of mathematics. By using the CELs, providing a balance in the delivery of instruction and assessment, and embedding the teacher expectancies, students will have a much better chance at success and excelling in mathematics. This just doesn't happen; thorough planning will increase the likelihood of success.

2

PREPARATION AFFECTS EVERYTHING

Whether you are in business, sports, or education, preparation matters! When watching your favorite team play, there is an expectation that the team is prepared for whatever the opposing offense or defense plans to throw at them.

We have all heard expressions describing the importance of preparation like, "Today's preparation determines tomorrow's achievements." To paraphrase former Dallas Cowboys coach Tom Landry on preparation: The only thing more important than the willingness to succeed and win is the willingness to *prepare* to succeed and win.

In most schools, there are good teachers and not-so-good teachers in rooms right next to each other. The suggestions offered in this chapter will have a tendency to level the playing field, with all teachers afforded the opportunity to share their knowledge of teaching with each other.

Most new teachers are brought into the profession using the "pier system." That is, new teachers are given a set of keys, a room assignment, information on where to find books and supplies, a roster, and a syllabus and then are sent onto the pier where they are thrown into the water to either sink or swim.

What is needed to be better prepared is a peer system. Can you imagine being a brand-new teacher and getting to meet your colleagues and have them share with you what you will be teaching, the order in which you teach it, where students traditionally have difficulty, resources and strategies to overcome those difficulties, what and how they test, how they grade, and actual materials to help you get started and learn your profession?

For this to occur, school administrators must also be prepared. Before the instructional school year begins, they need to identify goals and expectations, along with a plan to communicate and implement them, and follow up to ensure school personnel are receiving the assistance they need to optimize their performance in reaching those goals. These expectations have to be specific so it is perfectly clear what the supervisor will be looking for when observing and evaluating classroom instruction.

Preparation in mathematics is a great deal more than just knowing the chapter, section, and worksheet that will be used on a given day. Teachers having the least amount of success seem to be winging it, depending on memories from past performance. That's a problem, because the teachers are not learning and becoming better teachers.

To improve instruction, teachers must improve preparation and planning. The highest performing teachers know before instruction begins what they expect their students to know, recognize, and be able to do based on the common core standards, state standards, school district curriculum documents, high-stakes tests, and mathematical content. They also know how they are going to monitor their students' learning during a unit and assess it after the completion of the unit so student grades not only reflect their knowledge but also are both fair and portable.

Preparation and planning are keys to the success of teachers and their students. Preparing for a unit of instruction takes time and thought. The following might act as a guide on preparing for a unit.

Guide for Preparing for a Unit

1. Identify what students should know, recognize, and be able to do on the selected unit (specification sheet) based on the common core standards, state standards, district curriculum, and mathematical content.
2. Identify how long it should take to teach the selected unit (benchmarks).
3. Determine how and what to assess on the selected unit to help ensure consistency (portability) and fairness between classes of the same grade level or same subject (assessment blueprint).
4. Create a practice test using the specification sheet and assessment blueprint.

5. Using data and experience, identify topics within that selected unit in which students traditionally experience difficulty.
6. Identify ways to introduce concepts and skills to create interest and enthusiasm.
7. Identify linkages to previously learned math and outside experiences to review and reinforce concepts and skills and to address student deficiencies.
8. Identify simple, straightforward examples that clarify concepts and skills being introduced without getting bogged down in arithmetic.
9. Be able to visualize how student notes should be set up so they can study more effectively and efficiently.
10. Create algorithms/procedures based on concept development/ linkages/patterns.
11. Use choral recitation to embed definitions, algorithms, and formulas in short-term memory. Visually connect those to examples.
12. Create homework assignments that support instruction—more than just exercises.
13. Know how to close the unit by ensuring students see the big ideas and can differentiate between problems in the unit.
14. Create a strategy to monitor student learning and prepare students to be successful on unit/chapter tests.
15. Share with other teachers successful teaching strategies to overcome those difficulties or deficiencies.
16. Share content knowledge, resources, and expertise to address student success on the identified unit.
17. Using data, discuss ways to involve special education or English-language learner facilitators if specific student populations are not experiencing the same success as the general population.
18. Examine the results of the last unit test or other testing data to further determine strengths and weaknesses of individual students' understanding of subject matter.
19. Identify students not meeting proficiency on standards and a plan to remediate those students.
20. Identify instructional practices you will change for next year to correct deficiencies and improve student achievement.

Following a guide such as this will focus preparation on teaching and learning.

SPECIFICATION SHEETS

Before instruction on a major unit of study, grade-level or subject-area teachers should develop a *specification sheet*. That is, before they begin teaching, each group (third-grade teachers, algebra teachers, sixth-grade science teachers, etc.) should meet together and identify what they expect students to know, recognize, and be able to do and the timelines to accomplish those goals for a specific unit. This piece fits in very nicely to ensure all teachers are familiar with the common core state standards, school district curriculum guides, or already established benchmarks.

The types of questions being asked on a unit test should be based on the common core state standards, material being taught, and state and local standards. They should also reflect how that information might be tested on semester exams, criterion-referenced tests (CRTs), exit exams, National Assessment of Educational Progress (NAEP), and college entrance exams such as the ACT or SAT.

The following list is an example of a specification sheet for a unit on fractions. These are the things we expect the students to know, recognize, and be able to do.

- Definitions—fractions, proper, improper, mixed, equivalent, reciprocal
- Identifications—numerator and denominator
- Equivalent fractions—converting and simplifying
- +, –, ×, and ÷ fractions
- Borrowing/regrouping, whole and mixed numbers
- Algorithms for +, –, ×, and ÷
- Rules of divisibility: 2, 3, 4, 5, 6, 8, 9, 10
- GCF, LCM
- Common denominator—methods
- Draw models for =, +, –, ×, and ÷
- Ordering/comparing
- Applications (word problems)
- Open-ended concept or linkage

With experienced classroom teachers involved in this process, it might take fifteen or twenty minutes to create a specification sheet that is based on the common core state standards and school district curriculum documents.

BENCHMARKS

Following the guide above, setting time frames, or benchmarks, for teaching the material is the second major component in preparing for a unit. If one teacher determined it would take six weeks to cover a fractions unit and another indicated he needed only five weeks, that's okay. But, if one teacher indicated he needed two weeks and another scheduled twelve, then that is a problem that needs to be resolved. Chances are, if teachers are that far off on scheduling, then some are not covering the standards and others might not be addressing mastery.

ASSESSMENT BLUEPRINTS

The third major component in planning is the *assessment blueprint*. To ensure students are receiving balanced instruction, teachers should determine how they are going to assess their students before instruction begins. The assessment should have *balance*, as previously defined, and some type of agreement on the types of questions that promote *consistency*, *fairness*, and *portability* in the grading system. Portability means that a grade of B earned in one class would transfer and be equivalent to a grade of B in another teacher's classroom.

The blueprint does not identify specific questions, but rather the approximate number and type of questions that promote a balanced assessment. In math, for instance, sixth-grade math teachers might agree to have approximately twenty questions on a test, then determine how many might be computation, vocabulary/identification, concept/linkage, word problems, modeling, and so on.

Based on the specification sheet—the standards being assessed—an assessment blueprint should be constructed that identifies how the items on the specification sheet are to be tested and the number of items for each specification. The following is an example of an assessment blueprint for fractions.

- 2 definitions
- 1 identification
- 2 algorithms/information
- 1 rules of divisibility
- 2 concept/linkage problems—open ended
- 1 draw model
- 1 ordering
- 1 simplify
- 4 computation, +, −, × and ÷
- 1 GCF, LCM
- 3 word problems (applications)

It generally takes longer to come to consensus on the assessment blueprint. Teachers need to keep in mind that the assessment blueprint is a guide and teachers should work toward building a consensus; it is not a binding agreement. Classroom teachers continue to make up their own tests, unless they want to create common tests by grade level or subject. Fairness and portability of grades from class to class or between schools should be considered.

With respect to the blueprint, one teacher might decide to have four computation problems, while another might choose to have six. That is okay. The goal of the assessment blueprint is to assess students in similar ways and at approximately the same level of difficulty. This approach will help ensure portability of grades. It would not be fair to students, for example, if one teacher had a question like "Reduce $\frac{6}{8}$" while the teacher across the hall had their students reducing $\frac{111}{213}$ on the same unit test. It should be noted that the more the tests are identical, the higher the correlation between students' grades between classes will be.

The assessment blueprint gets into testing—teachers tend to argue strenuously over this. What we quickly realize is that teachers don't want their students tested on what they don't either stress or teach. For example, teachers who are constructivists may not want students to memorize facts or procedures. Other teachers might not see the value in students understanding what they are being taught and just want kids to memorize the rules. Experience suggests a heated argument will develop in these circumstances and hard feelings will follow unless a balanced curriculum is being followed. To settle such disputes, teachers should refer back to their specification sheet, curriculum documents, and state or common core standards to determine what should be tested.

PRACTICE TESTS

Construction of a practice test based on the specification sheet and assessment blueprint for a unit that contains items from the common core standards, state standards, school district curriculum, the math content in the unit, CRTs, NAEP, graduation tests, and college entrance exams such as the ACT and SAT suggests that the teacher possesses professional knowledge, is prepared, and will be in position to better ready his or her students for any test they must face.

The sample test in figure 2.1 reflects the items listed on the specification sheet and the number and types of questions on the assessment blueprint for a sixth-grade test on fractions.

This sample test was provided just as an example. A couple of things should be noted. First, the test did not exactly follow the assessment blueprint. It should also be noted that having teachers identify what they want their students to know, recognize, and be able to do is a straightforward process with which teachers readily agree. Second, the first part of the test, questions 1–6, did not involve any math computation or manipulation. It was strictly information students need to know to be successful on their test. This will be discussed more fully later.

Care should be taken to write test items so that students are exposed to the way in which those questions are phrased or tested on standardized exams. For example, in algebra, a direction on a teacher-made test might be to "solve" an equation. On college entrance exams, the same direction would be to "find the solution set over the real numbers such that . . . ." The way the question is asked might cause some students, especially struggling ones, not to connect what they learned in the classroom to what is being tested on high-stakes tests.

Fractions

For questions 1–3, write the definition for each.

1. Fraction
2. Proper fraction
3. Reciprocal
4. In the fraction $\frac{3}{8}$, the 8 is called the ________________.
5. List two methods for finding a common denominator.
6. Write the steps, as discussed in class, for adding fractions.

For questions 7–10, evaluate each expression. Simplify your answers.

7. $\frac{5}{7}+\frac{1}{3}$

8. $\begin{array}{r} 12\frac{1}{2} \\ -7\frac{2}{3} \\ \hline \end{array}$

9. $5\frac{1}{2}\times\frac{2}{3}$

10. $\frac{3}{4}\div\frac{1}{8}$

11. Find the LCM and GCF of 108 and 72.
12. Simplify the following fractions to lowest terms (simplest form).

    a. $\frac{8}{12}$ b. $\frac{27}{63}$ c. $\frac{111}{207}$

13. Write a five-digit numeral divisible by 2, 3, 4, 5, 6, 8, and 10, but not 9.
14. Order the following fractions from least to greatest. Show your work or explain the strategies that you used.

    $\frac{3}{4}, \frac{7}{10}, \frac{5}{7}$

15. If the numerator of a fraction remains constant and the denominator increases, what happens to the value of the fraction? (Assume the numerator and denominator are both positive.)
16. A student added $\frac{1}{7}+\frac{4}{7}$ with a result of $\frac{5}{14}$. The answer is incorrect. What is his error and how would you explain to him the reason behind the correct answer?
17. Draw a model to show that $\frac{1}{2}=\frac{4}{8}$.
18. Bob owns five-ninths of the stock in the family company. His sister Mary owns half as much stock as Bob. Jill owns the rest of the stock. What part of the stock does Jill own?
19. Joel worked $9\frac{1}{2}$ hours one week and 11 hours and 40 minutes the next week. How many more hours did he work the second week than the first?
20. A person has $29\frac{1}{2}$ yards of material available to make uniforms. Each uniform requires $\frac{3}{4}$ yard of material. How many uniforms can be made? How much material will be left over?

Figure 2.1. Model test for fractions

SHARING KNOWLEDGE AND EXPERTISE

After creating a specification sheet, establishing time frames, making an assessment blueprint, and constructing a practice test, experienced teachers should share their knowledge of where students traditionally experience difficulty on a particular unit. Rather than bemoaning the fact that students have performed poorly on those areas historically, teachers should exchange knowledge, resources, experiences, and successful teaching strategies with each other. Modifying instructional strategies and resources can result in greater student understanding and increased student achievement.

Teachers could increase their content knowledge by using this time to share their understanding of conceptual knowledge and application of the knowledge and skills taught in class. They might also examine areas in which the district or school has not performed up to expectation on state and national tests and address those areas of concern. Further, they might study their most recently administered test to determine strengths and weaknesses of their instruction. Once that has been accomplished, decisions can be made on how best to address weaknesses during the current school year and how instructional strategies might be changed in future years.

If specific student populations can be identified as doing poorly, the grade- or subject-level teachers might want to bring into their meetings English-language learner, special education, reading, or instructional strategists to recommend possible changes in instructional methodologies that would be beneficial to those identified students.

USING SIMPLE, STRAIGHTFORWARD EXAMPLES THAT WORK

Nothing ruins a good lesson like a bad example. Teachers who take great care in choosing their examples before instruction begins tend to have students with better understanding of the concepts and skills being taught and feel more comfortable in their knowledge.

To a layperson, the following two division problems look pretty much the same.

$$1355 \div 41 \text{ and } 1355 \div 47$$

But to more thoughtful math teachers, they realize while these two division problems may look alike, they are very different.

In the first example, the trial divisor works and there is no carrying. In the second example, the trial divisor does not work and there is carrying. When introducing long division, or any other concept or skill, successful teachers use simple, straightforward examples that work and clarify what they are teaching without bogging students down in arithmetic.

What that means to many students is the difference between success and failure. It is recommended that teachers identify the examples they will use in class in their planning to ensure they work and are not a variation of the problem that could negatively impact student learning and understanding. Build success on success.

As an example, in teaching long division, begin with example problems where the trial divisor works and there is no carrying. Once the students feel comfortable with that, move on to examples where the trial divisor works, but there is carrying. Finally, when students have mastered that, then use examples like the second example where the trial divisor does not work and there is carrying. That kind of scaffolding in instruction makes the students more comfortable and builds confidence in their ability to do the exercises they are assigned.

Using simple, straightforward examples that work and clarify what is being taught without bogging students down in arithmetic should be incorporated in almost all math lessons from basic math through calculus.

PREPARING FOR A UNIT TEST

One very effective way to improve student performance is to close out a unit by taking a couple of class periods to review what the students have learned by grouping key concepts in ways that make it easier for students to remember. Usually that means that the teacher compares and contrasts what was learned, so the kids see the big idea in a simpler context.

Many students find success in learning topics on a daily basis, then perform poorly on a unit test. The reason is that they don't see or understand how problems are similar and different. In first-year algebra, for instance, struggling students are typically taught five methods of factoring. As each is introduced, they get it and are able to do the problem. But when the test is administered, the results don't reflect their knowledge. While the students knew that $x^2 + 7x + 12$ and $6x^2 + 7x + 2$ were both trinomials—they looked the same—they didn't realize the method of factoring would be different, determined by the leading coefficient.

Another concern is an accepted fact by math teachers: that math is a language. Students must be taught vocabulary explicitly so they understand the questions being asked of them. They must also realize that the same question can be asked in different ways.

As can be seen, there is a great deal expected of classroom teachers. Better preparation and planning results in better instruction and increased student performance. The Components of an Effective Lesson provide a structure that helps teachers incorporate the teacher expectancies and prepare their instruction.

POINTS TO REMEMBER

At the very least, there should be hard evidence of preparation from each teacher or department that would include a specification sheet, an assessment blueprint, a practice test, the simple straightforward examples that will be used during initial instruction, a sampling of student notes, and homework assignments that support and reflect instruction that are more than just exercises.

The simple fact is that proper planning and preparation by teachers prevents poor performance by their students. As professionals, teachers need to realize where students might encounter difficulties and address those before they experience them.

We *can* improve preparation that results in increased student understanding and achievement. Let's do it!

3

INSTRUCTION REALLY MATTERS!

USE OF INSTRUCTIONAL TIME

State and local school districts usually determine the classroom time available to teachers and students. Regardless of the quantity of time allocated to classroom instruction, it is the classroom teacher and school administrator who determine the effectiveness of the time allotted. Teachers should strive to begin instruction immediately and be up either actively engaging students in instruction or walking around the room monitoring their understanding.

According to a survey conducted by the American Association of School Administrators, teachers identify student discipline as the single greatest factor that decreases time on task in the classroom. Generally, teachers with well managed and organized classrooms have fewer disciplinary problems. These classrooms typically have teachers who have established rules and procedures, are in the classroom when the students arrive, and begin class promptly. They reduce the "wear and tear" on both themselves and students by establishing procedures for makeup work, and they develop routines that increase overall efficiency.

The benefits of establishing these classroom procedures and routines become apparent as the total time on task approaches the allocated time. The research says: When teachers begin class immediately, students view them as better prepared, more organized and systematic in instruction, and better able to explain the material. Students also see these teachers as better classroom managers, friendlier, less punitive, more consistent and predictable, and as teachers who value student learning."

That's an awful lot coming from just starting class on time. But when you think of it and understand our culture, it does shed light on those conclusions. In countries such as Italy or Greece, it is all right for boys who are friends to hug or even kiss each other on the cheek. In the United States, hugging and kissing is not well received. The way American boys show affection is by hitting. So when two boys like each other, they show it by hitting their friend on the arm. If there was too much "love" being delivered in the punch, the other boy might respond with a little extra affection of his own. Too much of this affection might lead to a fight.

Routines—like beginning class immediately, reviewing recently taught material, orally reciting new material, having students take notes, and ending the class by reviewing important definitions, formulas, algorithms, and the daily objective—keep students engaged and on task. Quality time on task is not a "silver bullet" that can cure all the problems facing education; however, it can play an important role in increasing student achievement. Teachers must ensure that the entire class period is used to its full potential. That is, an academic focus and on-task behavior are [to be] maintained by the effective use of allocated instructional time.

MAKE A GOOD FAITH EFFORT TO TEACH THE ASSIGNED CURRICULUM

Just as you cannot spend a dollar twice, you can only use an hour once. Teaching bell to bell, optimizing the instructional period, is paramount to success. When not actively instructing, teachers should be walking the room to monitor student learning.

Before we get into discussion about actual lessons, there must be an understanding that teachers and administrators must do their job. For teachers, that means they should plan to make a good faith effort to cover their assigned curriculum. The district must ensure that the curriculum can be covered in the time allotted, and the school administration needs to ensure that content is being delivered in a way that increases student performance.

Most school districts have established policies similar to the following: Guides or course syllabi are established for all areas of the curriculum and are to serve as the basis for instruction in district. Members of a professional staff shall utilize these guides as a means of meeting the needs of individual students. Making a good-faith effort to teach the curriculum means that teachers plan to cover all the material in the appropriate syllabus.

The development of specific teaching techniques is the responsibility of the individual teacher. It is suggested that these teacher expectancies be incorporated into daily plans. These should be consistent with the district's objectives and proven principles of learning. In addition, many school districts have also established position statements or guidelines like the one in figure 3.1.

Imagine you are the parent of twins who are enrolled in the same school but don't have the same teachers. Would you expect them to be learning the same material at about the same time? Covering the established curriculum with the appropriate benchmarking will create the consistency needed to develop and maintain credibility in the community.

If teachers do not cover the curriculum assigned to them, students will end up with gaping holes in their knowledge. All too often, you hear teachers talking about how much time they must spend at the beginning of the year reviewing topics that students were supposed to have learned the year before. While I sympathize with those teachers and the work they think they must do to remediate those students, the fact is that by not covering their assigned curriculum, they will be contributing to the problem as well.

The first rule of being in a hole is to stop digging. To address these deficiencies, teachers could better utilize the idea of linking new concepts and skills to those areas of concern and reinforce those basic skills as they teach throughout the year.

STANDARD FOR QUALITY: Adopted secondary course syllabi serve as the basis for classroom instruction.

I. Instructional activities are correlated with stated objectives in adopted secondary course syllabi.
II. Resources are selected to support objectives in course syllabi.
III. Daily, unit, and semester planning includes goals and objectives contained in course syllabi.
IV. Appropriate accommodations and/or modifications are made in alignment with goals and objectives in adopted course syllabi to meet the instructional needs of all students.

Figure 3.1. Sample school district position statement

The simple fact of the matter is that teachers who spend too much time remediating students who have already been remediated will not be able to cover their own curriculum assignment, which will result in the school not showing growth. One of the first things that should be checked if a school is not reaching standards is whether they were teaching the curriculum assigned to them. How teachers teach is important, but what they teach is of greater importance! Teachers must teach the curriculum assigned to them.

BENCHMARKS

Is there a need for a consistent, standards-based curriculum? That question may best be answered by asking another question: Have you ever had students transfer into your class who have not acquired the necessary prerequisite skills and knowledge to be successful? If you answered yes, then you see the need for a curriculum that is consistent, not only between classrooms within a school but between schools as well.

Although curriculum guides, benchmarks, and syllabi provide classroom teachers with clear goals and expectations, they are often not accompanied by explicit time frames. Therefore, in order to maintain a consistent mathematics curriculum, benchmarks should be established. Benchmarks are approximate time lines by which particular concepts and skills are to be taught. It is suggested that teachers within a school teaching a particular class work jointly to develop benchmarks.

Once benchmarks are developed, they must be revisited on occasion to allow for necessary revision. Setting and following these benchmarks should ensure adequate coverage of essential course objectives that lead to mastery. By adopting these recommendations, these time frames would be established for each unit of study.

To further ensure students are meeting academic expectations, common periodic testing during the school year should be scheduled to determine the level at which students are achieving mastery on specific topics.

FRONTLOADING

We have discussed the importance of teaching the curriculum and using benchmarks to assure students are spending the appropriate amount of time on the concepts and skills being taught to reach mastery. Having said that, there are times when it appears the system is setting students up to fail. While teachers clearly want to teach to mastery, a dense curriculum can get in the way of achieving that goal. Teachers often talk to each other about the issue of *coverage versus mastery* and how time affects student performance.

Another factor that impacts student achievement is the sequencing of the material to be taught during the year. If important concepts and skills are left to the end of the school year, teachers might not get to them, or if they do, they might have to rush to cover the topic. Not having the time to address mastery will have a negative impact on student performance on high-stakes tests.

Frontloading suggests that teachers examine the curriculum assigned to them and determine the most important topics to be taught during the year. Once that determination is made, teachers should ensure they teach that material early enough in the year so they know they will get to it, teach to mastery, and have opportunities to review and reinforce those concepts and skills regularly during the school year.

Naturally, material should not be just arbitrarily moved around. Subskills still need to be taught and the foundation needs to be laid for what is to be taught. Let's look at an example.

Fractions are often taught in early grades. There is a clear consensus that students should be able to compute with fractions. If fractions were introduced late in the year, some teachers might not get to them, while others might feel rushed trying to cover them. The following year's teachers will be upset if their students come into their classes without the prerequisite knowledge and skills for them to teach the curriculum assigned to them. That would result in them trying to reteach the material, giving up the valuable time they need to cover the material they are assigned to teach.

COMPONENTS OF AN EFFECTIVE LESSON

Too often in life, decisions are made by default—by *not* making a decision. Many students actually make the decision not to go to college long before their freshman year of high school. The decision is made by their class selection—not by a conscious decision not to go. The same can be said of some classroom teachers: They do things without consciously making the decision, things that with a little extra thought they probably would have modified or done differently. Now that we have discussed protocols, expectations, and preparation, let's see how that affects the delivery of instruction.

The Components of an Effective Lesson (CELs) are a blueprint for classroom teachers to follow that is easily monitored by school administrators. If fully implemented, the CELs and teacher expectancies will result in increased student achievement. The CELs consist of the following:

- Introduction
- Daily review
- Stated daily learning objective
- Concept and skill development and application
- Practice (guided/independent/group)
- Homework assignments
- Closure
- Long-term memory review

Most teachers already loosely employ many of these components in their current lessons. These components are not controversial, but they are not typically being fully implemented. We will discuss each component in more detail below.

THE INTRODUCTION

As part of setting the stage for learning, students should be made aware of each day's objectives and how the topic might be used in everyday activities. Teachers should explain why students might want to learn what is being taught that day. The introduction provides classroom teachers an opportunity to create interest and enthusiasm in the lesson that will be taught. As an example, if the day's lesson is about the circumference of a circle, the teacher might explain how changing the tire size on a car will affect its odometer and speedometer readings and might reduce the life of the tires.

Everyone uses math in their daily life. Make those connections. With some concepts, teachers do a great job of introducing how the math they are teaching can be used. But we need to do a better job of introducing concepts and skills that create the connection between the real world and the math classroom more often.

DAILY REVIEW

Stop the warm-ups! Many teachers begin their classes with a warm-up activity that results in beginning the day's instruction with an exercise that results in a negative learning climate by asking students to do problems they don't know how to do. These warm-up activities also result in a loss of valuable instructional time. They tend to last anywhere from five to twenty minutes—time that cannot be afforded to waste.

All too often, teachers ask students questions during the warm-up, trying to engage students. Struggling students often don't know the answers. It aggravates the students and frustrates the teachers—a lose-lose situation.

For students who need remediation or reminders or who have been absent, the warm-up activity is a waste of their time and reinforces the idea they are not good in mathematics. The warm-up activity generally does not support the idea of building and using a success-on-success model.

Hollywood has figured it out; so should educators. When a television show is continued to the next week, the continuation show doesn't begin by asking the audience what they remembered—engaging as that might be. It starts with a quick, crisp, purposeful review of elements of part 1 that the audience needs to know to make sure they can follow, have some insight, and make sense of what is happening in the sequel.

Teachers should do the same. Rather than asking *engaging* questions with little to no response or providing an assignment that will keep kids busy so administrative tasks like taking attendance can be completed, struggling students would be better served if teachers immediately began their classes with a quick, crisp, purposeful review of recently taught material and the concepts or skills the students will need to recall to be successful in meeting the objective that day. The review could also include one or two representative samples from the previous night's homework assignment. That review refreshes student memories and sets them up for success for the day's lesson.

STATING THE DAY'S OBJECTIVE

To ensure the day's lesson is focused, teachers should state and write its objective on the board. The objective should be left on the board for the entire class period to remind the teacher and students what they are expected to learn.

An example of an objective, based on the circumference topic mentioned in the introduction, might be: "The student will learn the origin of pi and be able to find the circumference of a circle given the radius or diameter."

While many of us believe students know what we are teaching based on the presentation, it is surprising to know many students cannot identify the big idea being taught in specific lessons. Stating and writing the objective clearly resolves that problem. The objective should also be copied in the student notebook!

CONCEPTUAL DEVELOPMENT

Remember, it's not a matter of if students are going to forget information they have learned, it's a matter of when. Without concept development, students will not be able to reconstruct knowledge lost over time.

In mathematics classrooms that lack sufficient concept development, memorization of rules and algorithms is emphasized, but little or no attempt is made to help students understand the why of mathematical processes. Concept development should be as important as memorizing basic facts and algorithms. Students' understanding of, and comfort level with, new ideas is increased when concept development is done properly.

Sometimes students are able to get the right answer even though they don't necessarily understand the why. Mathematics then becomes an arbitrary set of isolated rules, which can often lead to future pitfalls. As mathematics becomes more abstract, "math anxiety" may develop if these rules and algorithms have not been developed with an understanding of why they work. Eventually, students can become frustrated and quit taking math, even if the grade they earned in their last class was average or above average.

Developing concepts or linking those ideas to students' prior experiences helps to explain the why and makes students more comfortable in their knowledge and understanding of mathematics. For example, rather than just having students "flip and multiply" when dividing fractions, the division algorithm might be developed through use of repeated subtraction as we discussed earlier. The Pythagorean Theorem might be explained by using the areas of the squares formed by the sides of a triangle.

Unfortunately, students all too often "tune out" teachers during concept development. Since students value what teachers test, concept development itself must be tested. Students might write a brief explanation of the development of a particular concept in their notes and as a part of the homework assignment and then be asked an open-ended question on a test where they must explain the origin of a rule or algorithm.

Questioning Strategies

Who's doing the talking in the classroom? For students to learn the language of mathematics, students have to read it, write it, hear it, and speak it. How questions are asked in a math classroom can either promote or hinder dialogue.

Questions can take different forms. The most common form of questioning in a math classroom is directed: "What's the answer to number 8?" A teacher might ask a student for an answer to a particular question and receive a one- or two-word response. Another form of questioning is regurgitation. Students might be asked for formula, procedure, or list. A third type of question might be referred to as cueing. In that situation, a teacher asks a question, and after the student responds, the teacher repeats the student response, then asks for more information by asking questions like, "Why?" And finally, there are conceptual questions—questions that have students explaining or linking a concept, comparing and contrasting. Conceptual questions elicit more than a one-word answer or list and cause students to communicate their thoughts and understandings of a particular topic.

All of these types of questions should be incorporated into daily lessons. To keep students engaged, teachers need to ask all four kinds of questions. Importantly, teachers need to allow students time to think of an answer before they just give it. Teachers clearly need to understand the value of "wait time" when asking students to engage in discussions.

To help teachers determine the types of questions they are asking of their students, administrators observing a lesson might list these four categories and determine the number and types of questions teachers are asking of their students. Knowing the emphasis of their questioning might help teachers get students more engaged in their learning.

Presentation Techniques

Nothing ruins a good lesson like a bad example. Teachers must take great care in choosing examples. Teachers need to be careful and pick simple, straightforward examples that clarify what they are teaching but don't bog kids down in arithmetic. Too many teachers think of examples as they are teaching, without much forethought, and wind up picking a variation of the concept or skill that results in confusion for the students. Before variations are discussed, it helps student understanding to first comprehend the big idea being discussed.

Building student confidence and building success on success goes a long way toward increasing student achievement. Introducing variations of the problems before student understanding is complete often distracts from learning the objective of the day. Use simple, straightforward examples that clarify what is being taught is referred to as "comprehensible input" in the world of teachers working with students whose second language is English.

Some teachers have difficulty keeping students focused on the lesson at hand. By "using thorough instructional planning and preparation" and "smooth, rapid transitions between activities," teachers can prevent a great deal of student inattention. Just as important is the recognition of different learning styles. Most secondary school teachers present their lessons visually and verbally while students sit in rows and work independently. This is in sharp contrast to elementary classrooms where students are customarily seated in groups and the majority of instruction is verbal and tactile.

Using a variety of teaching techniques should help develop mathematics concepts in a manner that focuses students' attention, increases their level of engagement, and decreases off-task behavior. When teachers use a combination of verbal and visual instruction, hands-on activities, demonstrations, guided practice, and both independent and group work (cooperative learning), they meet the needs of most students.

The integration of visual, auditory, oral, and tactile styles of teaching and learning ensures students of a better opportunity to learn and feel more confident and competent in their abilities to do and understand mathematics.

Visual Component

When learning new material, many students have a need to see it, or at least visualize it, before it becomes meaningful. This implies that teachers need to prepare lessons and materials well before actual instruction. This might include creating models, charts, slides, graphs, videos, manipulatives, overhead transparencies, or handouts.

Many times, the visual component of instruction coincides with the development or presentation of the lesson itself. That is, the teacher gives notes on the board or overhead while verbally explaining the material. This is a powerful instructional technique that many teachers use effectively. When teachers shortcut this process by only "talking through" multistep problems or having students mentally "visualize" complex situations, students may become confused and disinterested. Many students are primarily visual learners, and teachers must be concerned about and creative in meeting their needs.

Math classrooms should have a lot of board space to develop or link concepts and see patterns develop. Based on the development, students see patterns that lead to rules and procedures that should also be written on the board. And finally, problems should be done using those formulas, rules, or procedures. Using the board space in this way will help students as they practice using formulas and procedures.

If teachers are using an overhead projector, smart board, or PowerPoint presentation, this becomes almost impossible for students to do without interrupting the flow of the instruction because the teacher would have to continually go back and forth on the screens. Much more thought and consideration needs to be given in using available technology when teaching concepts and skills in math. Students should be able to visually scan the material quickly to increase their understanding.

Oral Component

Oral recitation is the practice of having the entire class recite important facts, identifications, definitions, formulas, algorithms, theorems, and rules during their initial presentation and later when these topics are revisited.

To assist students with remembering procedures, the best teachers have an example problem written beside the developed procedure so that, as students are reciting the procedure, the teacher can point to the specific part of the example that corresponds to that step, double-coding the information in the brain.

To ensure full participation, individuals are also called upon. The number of times an item needs to be repeated depends on the difficulty of the material and the ability of the class. To support the success-on-success model, have the entire class recite the new procedure by reading it off the board for approximately sixty seconds with an example beside it. After sixty seconds, erase it and have the students continue to recite. Now, to create the impression students are getting it, call on students who can recite it—not students who might not be getting it.

To enhance the motivation of students not fully participating, begin calling on students who can recite the procedure or formula on one side of the classroom, then begin to work in almost a straight line to the students who may not have been fully engaged. Their anxiety and attention might increase a little as they see students in their proximity being called upon to recite. And since those kids are being successful, they typically don't want to stand out as the only one not, and they will pay closer attention.

In real life, repetition is used frequently to aid memory or to make a point. For example, adults often use it to remember a license plate number or a grocery list. Advertisers use repetition in presenting their message to the public. How many times have you found yourself repeating a phrase from a commercial or humming a jingle?

Often, teachers report that students have performed poorly on assignments or tests because they did not study. While not a substitute for out-of-class studying, in-class oral drill and recitation provide an opportunity for important repetition, a tool in improving students' achievement. When used correctly, they compel students to become active participants in the lesson and teach them one method for memorizing new information. Oral recitation has the additional benefit of meeting the needs of auditory and English-language learners and teaches all students how to say, read, and write mathematics correctly.

Tactile Component

A National Council of Teachers of Mathematics (NCTM) motto states, "Math is not a spectator sport." The tactile learner learns by doing; he or she needs direct involvement in the mathematics process.

Teachers should provide tactile learners the opportunity to explore and understand mathematics through the use of manipulatives, hands-on activities, labs, group work, projects, and paper-and-pencil problem solving. Additionally, technology such as calculators and computers should be used to aid the tactile learner. When using these instructional aids and procedures, teachers may need to perform the activity as a learner prior to its presentation in order to fully understand the process involved and anticipate students' difficulties.

Practice (Guided/Group/Independent)

When people take up a new sport, they do not expect to be proficient immediately. One expects to practice a new activity to get better at it; long, hard, extensive practice is almost always necessary to become proficient. Learning mathematics skills can be equated to learning physical skills. Paced practice with frequent reinforcement, feedback, and evaluation is essential to master abstract concepts. To avoid shortchanging their students, teachers need to provide practice—guided, group, and independent.

As part of developing mathematics concepts, teachers need to give students opportunities to practice new skills with immediate feedback. Initially, teachers should include several examples as part of the explanation while giving notes. Guided and independent practice may be more than paper-and-pencil work. It may include labs, projects, and the use of technology.

Before students are sent home with homework (independent practice), guided practice should be extended to ensure that students are proceeding correctly. A couple of exercises, similar to the homework assignment, are provided for students to work on in class. Teachers should pace their students through guided practice to ensure time is not wasted. They should monitor students carefully, looking for points where they become hesitant, stuck, or confused.

If many students stumble on or fail to grasp a given idea or step in an algorithm, the teacher should immediately address the problem on a class-wide basis. If only a few students experience difficulty, these problems can be handled on an individual basis. In either case, a review of the guided practice exercise is recommended before students leave class to begin independent practice.

It must be noted that guided practice should not be "starting homework." Students frequently dawdle during or entirely misuse class time allocated to practice if the time is not paced or for "homework." Homework, for the most part, is to be done outside class time; guided practice is done in class with immediate feedback. As an incentive, teachers may add an assessment component to guided practice—essentially a participation grade.

Allowing students to work in groups provides students an opportunity to explain concepts and skills, to use the language. That communication has the benefit of getting students thinking, reflecting, and organizing their thoughts, which not only helps other students in understanding a concept or skill but also clarifies the student's own understanding.

Closure

At the close of the lesson, the daily objective should be repeated by the teacher or one or more students to emphasize what was taught that day. Teachers might also ask their students to summarize orally or in writing what they learned, how the lesson related to previous knowledge, and how it might be used. By listening to their students, teachers would have a better idea if students understood the lesson. It's just another way to monitor student learning.

Long-Term Review

A second review should be employed whenever possible at the very end of the instructional period to address long-term knowledge, student deficiencies, and mastery and to prepare students for high-stakes tests. The reviews should also be based on student performance—not a whim. The long-term review will often address concepts and skills that were learned previously and may not be part of the current year's curriculum.

High-stakes tests require students to maintain knowledge over time, and without these long-term reviews, there is no doubt students will forget information over time. Forgetting in school is often translated as either not being taught or not having learned. In either case, it's not only a problem for the students; it's also a public relations nightmare for the schools. Schools and math departments must have a plan to address the long-term knowledge of their students. Allowing students to forget information over time is not a viable option.

If the second (long-term) review is employed during the last five to seven minutes of a class, students will remain on task until the end of the allocated time. That will maximize time on task as well as having the additional benefit of cutting down on potential discipline problems. Implementing a second review period also has the added benefit of providing teachers more opportunities to address mastery and deficiencies and to prepare students for college entrance examinations such as the ACT and SAT or high school exit exams.

Employing the first review of recently taught material, using linking to introduce new concepts and skills, and making use of a long-term memory review whenever possible will give students with higher than normal rates of absenteeism a greater probability of being able to follow that day's instruction. That is much better than losing another day of instruction by not understanding or being able to follow the lesson, which oftentimes leads to student frustration and discipline issues.

INAPPROPRIATE TECHNOLOGY

Have you ever walked into a kindergarten and noticed that the classroom desks and chairs are very small and that there is normally a bathroom with smaller toilets sitting pretty close to the ground? If so, you probably made an assumption that the room was constructed that way because of the purpose of the room and the size of the children using it.

If you look at math classrooms around the country, you might notice they have a particular construction as well. The most obvious is the number of whiteboards or chalkboards in the room. In a typical math classroom, rarely would you see just two large boards—there are typically three or four. As there is a reason for having smaller furniture, bookshelves, and bathrooms in kindergarten, there is a reason for having multiple boards in a math classroom.

When math teachers introduce new concepts, they usually want to draw a picture and be able to refer back to the picture as they work through the problem. That takes board space. If the teacher is trying to develop an idea or concept by having the students examine problems looking for a pattern, that takes a lot of board space so students can look back at the problems to discover the pattern. After the picture is drawn, examples are placed on the board, then a rule or procedure is developed. All should be visible to the student.

Schools are now investing hundreds of thousands of dollars in new technology, often referred to as intelliboards or smart boards. These boards, like the overhead projector and PowerPoint presentation, do not allow students the opportunity to see the concept or pattern developing or to see the whole picture from the start to finish because they have to go on to the next screen to finish the work. The inappropriate incorporation of this technology can fly in the face of what is described in the research as best practices.

Like all technology, there are appropriate times to use this technology. Placing the smart board right in the middle of connecting whiteboards is really bad planning. The smart board should be placed where it can be used most appropriately and not take away from good teaching. More serious thought needs to be given before buying technology to ensure it supports best practices.

Many products, programs, and services are purchased for the "wow!" factor—mostly because someone they know purchased it. It gives the appearance that progress is being made. But the reality is, and always has been, that *what works is work!* Buying the razzle-dazzle, glitzy products with blinking lights, horns, and whistles doesn't help students learn or result in increased student achievement. Too many educators are mistaking activity or having the most up-to-date toy for achievement. In fact, some administrators are so enamored with the showmanship associated with the new technology that they don't seem to realize when it is masking poor instruction.

Whether math teachers are using a stick in the dirt, chalk, or smart boards, or are learning online using their smartphone, tablet, or computer, students still have to see patterns develop for understanding. They still have to learn the language of math and practice to gain procedural fluency. They still have to study and memorize information. Technology is not a silver bullet.

COMPONENTS OF AN EFFECTIVE LESSON: SUMMARY

Before presenting a lesson, refer to the specification sheet and assessment blueprint for the unit.

Introduction

- Set the stage for today's lesson (why am I learning this, students will take notes, participate in a group activity, etc.).

Daily review

- Provide review for short-term memory on recently taught material.
- When correcting homework, provide immediate and meaningful feedback and hold students accountable.
- Keep reviews and homework checks brief.

Daily objective

- State and write the objective before introducing the day's main lesson and have students record this in their notebooks.

Concept and skill development and application

- Teach the big concepts.
- Provide the "why" for rules and algorithms.
- Link concepts to previously learned material and/or real-world experiences.

- Utilize a variety of techniques: Students need to see it, hear it, say it, and do it
- Hold students accountable for taking notes and keeping mathematics notebooks.

Guided/independent/group practice

- Practice can be done at different times throughout the lesson to help students process information.
- Students need time to think, analyze, work on problems, discuss their solutions, and become problem solvers instead of watching the teacher do all the work.
- Practice can be done as an entire lesson that enhances conceptual understanding and/or application of concepts through inquiry, investigation, discovery, lab, or problem-solving activities.

Homework assignments

- Assignments should consist of what teachers value and include a variety of assessment items, including definitions, computations, explanations, applications, etc. (see the assessment blueprint for the unit).

Closure

- Have students explain in writing what they have learned and apply it.
- Restate what was taught.

Long-term memory review

- Address mastery, student deficiencies, and high-stakes tests, and stress important ideas—not necessarily part of this year's curriculum, but based on student knowledge.

POINTS TO REMEMBER

The old axiom is true: Testing drives instruction. So, to ensure proper preparation and instruction, school administrators should ask to see teachers' practice tests before instruction begins as evidence of preparation and to ensure the curriculum is being followed, benchmarks are being adhered to, the tests are fair, and grades are portable between teachers.

By making the test in advance, teachers will be better able to identify areas in which students traditionally experience difficulty and provide resources and strategies that assist students in their learning. The test must be balanced and contain the types of questions that students will see on high-stakes tests such as CRTs, NAEP, semester exams, exit exams, or college prep exams.

Not only should teachers be well prepared, but administrators must also reexamine how they supervise and evaluate instruction. If a school administrator is observing classroom instruction and can't follow the day's lesson, then he or she should be questioning how the students in the class are understanding it. This is not meant to imply that principals should be able to teach the lesson or even know if something is being taught incorrectly; rather, what it suggests is that school administrators should be able to follow the lesson and it should make sense to them.

If administrators cannot follow the lesson, then some questions need to be asked of the teacher during the post-observation conference about what they can do to more fully and appropriately develop the concepts or skills so students are more comfortable in their knowledge and understanding, resulting in increased student achievement. That means better preparation and planning. And finally, school administrators should ensure teachers are incorporating memory techniques in their instruction to assist students in their learning.

We can improve delivery of instruction to enhance students' understanding and help them remember information over time. Let's do it!

4

WRITE IT DOWN

When asked, memory researchers reported that the number-one "memory aid" they themselves use is "Write it down." In school, writing it down is called note-taking. Student notes should reflect and support instruction. If student notes are nothing but examples, one must question what the teacher did in the form of instruction.

Through proper preparation, teachers should be able to clearly visualize the notes students should be taking based on their instruction. If teachers are not presenting material in an organized fashion, a manner that would help students be more successful learners, and if students are left to their own devices to take notes without specific guidance from their teachers, then we are setting our struggling students up to fail.

Note-taking is important for a number of reasons. It keeps students engaged in learning, which keeps students busy and results in fewer discipline problems. The notes students take in class should reflect what was taught in that lesson. They should help students complete their homework assignment and serve as a foundation to prepare for unit or chapter tests and to review for high-stakes exams such as end-of-year tests, college entrance exams, or exit exams.

The notes shown in figure 4.1 reflect what is most often seen in a math classroom: a rule accompanied by a number of examples. While initially learning, students might be able to follow that instruction, record this information, and get many of the problems correct on their homework assignment. But, these types of notes will result in student difficulty. Not having

When you multiply exponentials with the same base, you add the exponents.

Ex. 1	$3^3 \times 3^5 = 3^8$
Ex. 2	$5^4 \times 5^6 = 5^{10}$
Ex. 3	$3^2 \times 3 \times 3^4 = 3^7$
Ex. 4	$5^2 \times 7^3 \times 5^4 \times 7^5 = 5^6 \times 7^8$

Figure 4.1. Example-based note-taking

the reason behind the rule will result in students thinking mathematics is like "mathemagic"—particularly when this concept is presented more abstractly in an algebra class.

Another concern with notes that look like this is that students will fall into very predictable traps. For instance, in example 3, the second number did not have an exponent. Many students, without conceptual understanding, will tend to believe that since there is no exponent, the exponent is zero.

Notebooks in many math classrooms typically record nothing more than example after example problem. However, complete notes should include a date, title, objective, vocabulary and notation, identifications, English–math translations, conceptual or pattern development that leads to a rule, the rule, example problems using that rule, and explanations that would help students understand the concept when something out of the ordinary occurs or when they review at a later date.

Teachers should assist students in setting up their notebooks to avoid visual overload when trying to study. Classroom teachers need to remember that at the end of the school year, textbooks are often collected, leaving students with only one resource to review and reinforce what they have learned—their notebook. Student notes are important! Student notes should reflect and support instruction.

Teachers should require students to take notes in all mathematics classes. Notebooks keep students engaged in learning, help them complete their daily homework assignments, enhance their study, and act as a foundation from which to prepare for tests, both unit and high-stakes tests. Also, since students are not allowed to keep their textbooks, the student notebook is usually the only mechanism available for review in later years.

Note-taking is a process used by students to record important information that they are trying to understand and need to remember. Because of the importance of a student notebook, teachers need to be prescriptive and directive in how notes are taken and accommodating in their instruction so students can take notes.

Student notes should reflect instruction and what was written on the board. Notes should typically include a title, the date they were taken, the daily objective, definitions, identifications, English to math translations, pattern or concept development with pictures or diagrams that leads to some conjecture, a formalized rule or algorithm, and a number of example problems used in guided practice.

Teachers should also encourage students to write an explanation of what led to the procedure being used to manipulate or solve problems. Explanations in notebooks are especially important when a problem-solving method might be construed as a "trick" and whose rationale would not be immediately obvious to the student when reviewed at some future date. The "notes" in figure 4.2 will increase student learning.

These notes are a great deal more extensive than the notes shown in figure 4.1 with just a rule and sample problems. If students were to revisit these more complete notes over time, the math would make much more sense to them.

Finally, while note-taking is a student responsibility, teachers need to hold students accountable for taking notes. This need not be complicated or time consuming, but it must be done frequently and consistently to further encourage students to take notes. One idea used by many teachers is to have notebook quizzes. Those quizzes consist of asking students to open their notebook to a certain date and write the definition or procedure given on a particular topic. That translates to free points if they took and kept the notes that should be required.

Notes are a very important component in increasing student achievement. The notes will help the students complete their homework assignments and should be the primary vehicle used by students to prepare for tests.

It is recommended that teachers be very prescriptive and directive, telling the students not only what to write down, but where to write it in their notebook. Spacing is important in student note-taking. If notes are crunched together, it will result in visual overload. That overload won't help students study more effectively and efficiently. A student's notebook sitting in the first row, first seat, should be almost identical to the student sitting in the fifth row, fifth seat, if they are following teacher directions.

Exponentials

March 16, 201X

Objective—Students will be able to define an exponent, identify the the base and exponent, evaluate exponentials in standard form and multiply exponentials.

2^3 exponent
base

In the number 2^3, read two to the third power or two cubed, the three is called the exponent and the two is called the *base.*

Exponent—an exponent tell you how many times to write the base as a factor.

Ex. 1 $4^2 = 4 \times 4$
Ex. 2 $5^4 = 5 \times 5 \times 5 \times 5$
Ex. 3 $10^6 = 10 \times 10 \times 10 \times 10 \times 10 \times 10$

Evaluating Exponentials in Standard Form

Ex. 1 $4^2 = 4 \times 4 = 16$
Ex. 2 $5^4 = 5 \times 5 \times 5 \times 5 = 625$
Ex. 3 $10^6 = 10 \times 10 \times 10 \times 10 \times 10 \times 10 = 1{,}000{,}000$

Exponentials with base 10 are easily evaluated by examining the following pattern.

$10^1 = 10$ $10^2 = 100$ $10^3 = 1{,}000$ $10^4 = 10{,}000$ $10^5 = 100{,}000$

The pattern suggests on an exponential with base 10, the exponent determines the number of zeros.

Exponentials are a shorthand method (mathematical notation) to express very large or very small numbers and can make computation easier.

In everyday life, exponentials are expressed using shorthand.

$4M translates to $4 million or 4×10^6
$5B translates to $5 billion or 5×10^9
$6T translates to $6 trillion or 6×10^{12}

Multiplying Exponentials with the SAME Base

To simplify products of exponentials with the SAME base, use the definition to write them as factors.

Ex. 1 $2^3 \times 2^4 = (2 \times 2 \times 2) \times (2 \times 2 \times 2 \times 2)$
How many times are we multiplying 2? Answer 7.
Therefore, $2^3 \times 2^4 = 2^7$

Ex. 2 $3^3 \times 3^2 = (3 \times 3 \times 3) \times (3 \times 3)$
How many times are we multiplying 3? Answer 5
Therefore, $3^3 \times 3^2 = 3^5$

Figure 4.2. Complete class notes

Ex. 3 $5^3 \times 5 = (5 \times 5 \times 5) \times 5$
How many times are we multiplying 5? Answer 4
Therefore, $5^3 \times 5 = 5^4$

Looking at those three problems with just their answers, do you see a pattern that would allow you to simplify these problems in your head?
Ex. 4 Try to simplify $3^{20} \times 3^{45}$ in your head.

It appears that when you multiply exponentials with the SAME base, you merely add the exponents. That pattern would suggest the following rule:

Rule 1. When you multiply exponentials with the same base, you add the exponents.

Use Rule 1 to simplify the following products:
1) $4^5 \times 4^7$
2) $7^3 \times 7^6$
3) $2^{10} \times 2^{35}$
4) 11×11^9

N.B. Notice in #4, there was no exponent written with the first factor, if you did the problem by definition the answer would be 11^{10}. That suggests when a number does not have an exponent, it is understood to be ONE!

Figure 4.2. ***Continued***

There are times when it is appropriate to ask students to put their pencils down and to watch and listen without taking notes. Some concepts and patterns are not as easy to see as others. Knowing this, to develop a concept or identify a pattern that is not easily recognizable or is more complex, it might be wise to ask students to just watch as the idea is being developed without the distraction of them writing. After a few problems have been demonstrated to help them see the pattern develop, then go back over the same development while the students take notes.

Pacing students through the note-taking process is also important to conserve instructional time. Students will take as much time as they are given to write their notes. Give the students enough time to take the notes, but move them along.

Note-taking should be part of the learning process, not done in isolation at the beginning of the period to be referred to later. They should be given in context with the lesson. If students have an Individualized Education Program that requires notes be provided, then provide those students notes. Otherwise, the students should be taking notes as the lesson is being taught.

STAR SYSTEM

When working with students who have not had a great deal of success in mathematics, implementing the star system helps tremendously. The star system is nothing more than a highlighting system that teachers typically employ in their classrooms. If teachers don't use a formal highlighting system, they typically use an informal one by saying things like "This will be on the test!"

The star system creates additional emphasis on topics to be tested—the number of stars do not reflect rigor. For instance, by placing three stars (***) next to something in their notes based on the instruction, that means that item will be in the notes, homework, practice test, and real test—unchanged. Three-star questions do not contain any computations or mathematical manipulations. They are questions that reflect what students need to know to successfully answer other questions on the test.

On an algebra test, three-star questions include definitions, identifications, word translations, algorithms, or formulas such as the quadratic formula or the formula for finding the discriminant or vertex. In geometry, three-star questions could include theorems or postulates. Again, these three-star questions should reflect information students need to know to be successful. Three-star questions are on the test unchanged from the notes, homework, and practice test.

Placing two stars (**) by information in the notes means that that type of problem will also be on the test, maybe with numbers being changed or a minor modification in the problem. As an example of a two-star problem, the notes could contain a problem involving finding the distance between two points. On the homework and practice test, different numbers might be used. Two-star problems are typically the types of problems assigned as exercises on the homework.

A one-star (*) problem highlighted in the notes suggests this item or a related one could be on the homework or practice test. As an example, students would be directed to place one star by the *procedures* for adding, multiplying, and dividing fractions. Only one of the procedures would be tested. As another example of a one-star problem, students might be asked to find the midpoint of a line segment or, given one endpoint and the midpoint, be asked to find the other endpoint. Those questions would be assigned for homework, and one of them might appear on the practice test. But only one would be on the real test.

Three- and two-star questions are those that students will know are on a unit test. The one-star questions allow teachers to ask broader questions so their curriculum is not constricted.

POINTS TO REMEMBER

As administrators are monitoring instruction, they should examine student notebooks. If the notes being taken by students do not reflect and reinforce the daily instruction—if they don't include a title, the date they were taken, the daily objective, definitions, identifications, translations, pattern or concept development with pictures or diagrams that leads to some conjecture, a formalized rule or algorithm, and a number of example problems used in guided practice—then they will not help them complete their daily homework assignment or prepare for a unit test.

A school administrator observing that should sit down with the classroom teacher and recommend changes to enhance the educational experiences of the students in that class. The administrator might suggest that the teacher help out the students having white space by being more directive when giving student notes. It may be necessary that more examples be provided by the teacher so the students can see a pattern emerging that leads to a rule. If the teachers were using an overhead, a smart board, or PowerPoint, the administrator might recommend that a chalkboard be used so the students can see the pattern develop in its entirety.

Notes are important for student success. School administrators overseeing math instruction should ensure student notes truly assist students in learning and reflect instruction. The star system highlights and focuses student attention on what teachers want them to know, recognize, and be able to do. Note-taking is essential in leading to student success.

We can help struggling students take better notes that will help them study more effectively and efficiently, resulting in increased student performance. Let's do it!

5

MAKING HOMEWORK ASSIGNMENTS WORTHWHILE

Homework is more than just assigning exercises. It is just one of the ways student learning is monitored and enhanced by highly effective classroom teachers.

"I couldn't do it! I didn't understand it!" How many times have teachers heard that as the reason students did not do or complete their homework assignments? For too many teachers, it is a daily occurrence that results in a great deal of instructional time being wasted and frustration on the part of both students and teachers. By rethinking homework assignments, classroom teachers can take those excuses off the table and provide a homework assignment that supports instruction and encourages study. Well-prepared teachers clearly make a difference.

As educators, we ask parents to set aside quiet time every night for their child's study and to ensure the homework is completed. Many parents do as requested, only to find that their kids don't know how to do the exercises. The following recommendations will help students study by connecting the daily instruction to notes and homework assignments. Parents should be able to look at notes given that day reflecting the instruction and see the pattern development, algorithms developed from that conceptual development, and the examples that will better enable them to help their children complete the homework assignment.

Student achievement rises significantly when teachers regularly assign—and students regularly complete—homework. The additional study that homework provides benefits students at all ability levels. Furthermore, homework gives students experience in following directions, making judgments and comparisons, raising additional questions for study, and developing responsibility and self-discipline. While homework is not the only or the most important ingredient for learning, achievement is often diminished in its absence. For all those reasons, homework must be assigned with the expectation it will be completed.

To maximize the positive benefits of homework, teachers need to give the same care in preparing homework assignments that they give to classroom instruction. Preparation matters. Teachers must carefully prepare the assignment, thoroughly explain it, and give timely comments when the work is completed. Homework should reinforce what was taught and learned. Students should not be assigned exercises if they are unsure of how to do the problems.

Homework should reflect what teachers' value—what we have defined as balance. Besides assigning a set of exercises for homework, teachers should also assign reading and require students to copy definitions, identifications, formulas, and algorithms and to write brief explanations of the day's work as part of their homework assignments. As students become accustomed to seeing these items as part of notes, homework assignments, and tests, they will also begin to understand their value. And don't forget to assign reading as part of the assignment.

Too often, homework assignments in middle school and high school read like this: "Page 89, Problems 1–33, odd." A good number of students then return to school the next day complaining to their teachers that they could not do the homework assignment because they did not understand how to do it.

Students would be better served if the homework reflected what teachers wanted students to know, recognize, and be able to do, as in a homework assignment that reads like the following:

- Read Sec. 2.3, "Solving Linear Equations"
- Write the Order of Operations
- Write the strategy for solving linear equations
- Explain how the Order of Operations is related to solving linear equations
- Page 89, Problems 1, 3–5, 7, 11, 18–21, 27, 31, 33

The second example incorporates into the homework assignment what math teachers say they value: vocabulary and notation, conceptual understanding, procedures, reading and writing, and practice exercises.

If a middle school homework assignment was based on a fractions unit, the homework assignment should look more like this:

- Read Sec 3.3, "Add/Subtract Fractions"
- Draw a model to add $\frac{1}{3} + \frac{1}{2}$
- Write the procedure for add/sub fraction
- Why don't you add the denominators when add/sub fractions?
- Name two ways of finding a CD and under what circumstances would you use each method?
- Page 162, Problems 1, 2, 4, 5, 12–16, 23, 27, 31, 33

While having students practice is important, a homework assignment should not be about just completing exercises. It should also be designed to help students learn—to encourage them to study. A homework assignment like this would better reinforce what is being taught in the class and recorded in the notebook than just providing a page number and problem set.

The simple fact of the matter is that in the more complete and thought-out middle school assignment, students who have to write out the procedure for adding fractions would be much more likely to be able to add the fractions that were in the exercises. The same would be true in the algebra assignment. If the students are asked to write the strategy for solving linear equations, the probability of them being successful with the exercises increases dramatically.

The real good news is that the first four or five questions on a homework assignment came directly from the instruction that was recorded in the student notebooks. That means students can go directly to their notes and answer questions like "draw a model," "write a procedure," and so forth, because that information is directly answered in their notes. And, if students can answer *those* questions, they are much more likely to be able to complete the homework assignment.

Homework assignments such as these being recommended have the added advantage of taking the excuse away of "I couldn't do it," because the first set of questions, before the exercises, can be directly answered by reviewing the daily notes.

Also, if you were working on a concept or skill for multiple days, some of those definitions, procedures, or formulas can be asked on multiple homework assignments to ensure the students are learning them. And, if the teacher thinks it might benefit the students to write that information more than once on a given day, that can be assigned also. Repetition is important to student learning.

When students first learn a procedure, they generally have to "think/remember" their way through the process. After initial practice, the procedure should become almost automatically applied. If it does not, teachers may want to consider giving more practice exercises.

A common trap teachers fall into is going over each problem that was assigned for homework the next day. That can result in a monumental waste of instructional time. A better way would be to incorporate two or three of those exercises from the previous night's homework in the first review so you can address potential questions and maximize instructional time. Teachers could then ask students to look at their solutions and see how the quick, crisp, purposely review of recently learned material, including homework exercises, addressed any of their concerns.

STUDYING

If we listened to students carefully, we would find that students "do" homework and "study" for tests. Many teachers assume that when they give homework each night, kids are studying. But, when you ask students what they do when they get the last question answered on their homework assignment, they will tell you they close the book and go on to the next subject. They simply do not equate completing a homework assignment with studying. That could change if the homework assignments given by teachers followed the recommendations provided.

If you want your students to be successful, teach them how to learn, and provide assignments that reinforce that. Many students believe the reason some kids are successful in math is because they are smart. They don't equate studying with being successful or being smart. Talk to your students about how they learn: Are they visual, audio, or kinesthetic learners, or a combination?

Most students don't know how they learn. Students who really don't know what it means to study use strategies such as the "stare and glare" method. Others use the "pray to God" method, and still others, because of an after-school job tend to grasp onto the "osmosis" method: They place the open text on their chests, take a nap, and hope that when they awaken somehow that print will have worked its way to their brain. While none of these methods are very effective, students continue to use them because they don't have other strategies that work any better for them.

Some students need to be taught how to study effectively and efficiently. They need to know that methods like saying it, writing it, and having someone ask questions—like they did in primary school—still work. Poor students tend to study until they "almost" have the information, then quit. Students need to be taught that after they study, they should be able to discuss what they studied without having their notes as a prompt—or else they need more study. Sometimes teachers inadvertently fail their students by not requiring them to verbalize the knowledge they gained.

Bottom line, teachers need to reinforce to their students that accomplishment is more dependent on hard work and self-discipline than on innate ability. School administrators also need to be told the same thing: Adoption of the latest new program will not lead to success, *work will*. In fact, the only place *success* comes before *work* is in the dictionary.

We also need to talk to students about their concentration times. How long can they study before they start looking around? I know I'm good for about forty-five minutes. While I can talk for days, after forty-five minutes of listening to someone else, I begin to notice how many lights are in the ceiling, a bird flying by the window, and a whole host of other items. Students need to know not only how they study but also how long they can study effectively and efficiently. Students need to be taught to extend their concentration time by studying a few more minutes before they take a short break from studying.

How many times have you heard frustrated parents tell someone they sent their children to their room for two hours to study to improve their grades? If a student whose concentration time is only thirty minutes is sent to his room for two hours to study, chances are great that an hour and a half will be wasted. Students need to be taught how to study effectively and efficiently. If teachers don't tell them, who will?

For students to be successful, classroom teachers need to be much more explicit about their expectations of students. Telling them to "study" just is not enough. Studying includes reading, thinking, reflecting, organizing, writing, analyzing, visualizing, reviewing, remembering, recalling, and more.

Studying needs to be explained to them, and probably modeled. Students should be told that while they are reading their notes, they should be able to visualize what they are reading. Furthermore, they should pause and think, reflect, visualize, and organize that information in their head. They should also know that as they are going over examples, they should be able to go to the next step without looking at their notebook. And after they have read those notes multiple times, that's when, with a closed notebook, they should recall what is in the notes to help them complete their homework assignment.

Notice that there is reading and writing in the homework assignment. Most teachers believe that math is a language, and language cannot be learned effectively without reading, writing, and speaking it.

READING

A common complaint among secondary teachers is that their students can't read at grade level or read their textbook. Yet today's high-stakes testing is composed of quite a bit of reading and problem solving. Secondary subject-specialist teachers are often in denial with respect to their role in teaching students to read. The vast majority of students have been taught to read by using fiction texts. The way students read fiction is different from the way technical material is read. Math students have to read differently to understand their math.

Who's going to tell this to the students if it's not their math teachers? It appears that the way most secondary math teachers are handling the reading problem is by ignoring it. In fact, a common strategy seems to be to stop giving reading assignments because the kids don't understand them. It really sounds stupid when you say that out loud. Not assigning reading will surely exacerbate the problem and continue the downward spiral.

Understanding a problem is surely part of mathematics. But you can't understand problems without having acquired vocabulary and notation, and that can't be done without reading. Math teachers have the responsibility to help their students read their texts. Reading a math text is different from reading a novel. Students and their parents might not realize that, so classroom teachers have to teach students to read mathematics.

Imagine studying the following theorem: *In a right triangle, the altitude drawn from the right angle to the hypotenuse divides the hypotenuse into two segments. The length of the altitude is the geometric mean of the lengths of the two segments.* Notice all the terminology: *right triangle*, *altitude*, *hypotenuse*, *segments*, *geometric mean*. A student who is not comfortable with that language might see learning that theorem as somewhat difficult—especially if he or she cannot visualize the theorem.

The math is easy; it's just solving for a variable in a proportion. So, just as students would know to stop on a red light, they should know that a geometric mean is nothing more than setting two ratios equal (proportion), where the means represent the same number.

$$\frac{4}{10} = \frac{10}{25}$$

In our theorem, the altitude is the geometric mean of the length of segment 1 and segment 2. Typically, one of those lengths would be unknown, and we would solve for it.

$$\frac{segment1}{altitude} = \frac{altitude}{segment2}$$

Teachers talking through this theorem, without drawing a picture, would give students little or no idea what this theorem is saying—just like the non-math teachers reading this. The visual is important to student understanding.

In terms of everyday study, and especially for preparing for a test, teachers need to make students very aware of the terminology they will encounter and how that terminology translates to simple math concepts and skills. When assigning reading in mathematics, teachers should explicitly introduce new vocabulary and notation before assigning the reading. Teachers should preview the reading and connect the reading to previous knowledge. After the students have read the assignment, teachers should check for student understanding of the reading and correct any misunderstandings—just as they do with homework problems.

Students should have paper and pencil to assist them in their reading of math content. Students reading mathematics don't read by chapter, section, page, paragraph, or sentence; they typically read phrase by phrase—every word counts. Students should copy important information, definitions, formulas, examples, and pictures to help them comprehend what they are reading.

When students read a novel, their eyes tend to follow the print from left to right at a constant rate. It's not the same when reading a math text. As students read an assignment, their eyes will dart back and forth from their reading to diagrams, examples, diagrams, examples, and back to their reading. They will generally reread a phrase a number of times before they feel comfortable enough to continue reading.

Students who haven't experienced success in mathematics don't like or feel comfortable reading their math text. They want it explained to them. Such students see reading a math textbook as futile, a waste of their time, and intimidating. Teachers need to teach students how to successfully read their math text. If the math teachers don't, who will?

Teachers cannot increase student achievement in mathematics if the students cannot read mathematics—if they cannot translate English to math and math to English. All of today's high-stakes tests are made up of word problems, so students have to know how to read mathematics.

Site administrators should ensure that their staff is teaching students to read in the content areas. As administrators observe instruction, they should see evidence of reading assignments, see the new vocabulary and notation introduced explicitly, and hear teachers previewing the reading, connecting the reading to previous experiences, checking for understanding, and correcting their understanding of what has already been read. If school administrators are not checking for this teacher expectancy, they should not expect it to be happening in the classroom—nor should they expect an increase in student achievement.

WRITING

Writing helps students clarify and solidify what they have learned and helps them respond to what they have read. Teacher expectancies such as emphasizing vocabulary, reading, and writing are seen as important because of their connection to language acquisition. Educational researchers have identified vocabulary as the single most important factor that leads to comprehension—student understanding.

Classroom teachers should incorporate a number of writing tasks into their daily instruction. Students might be asked to explain a concept, write a word problem, illustrate a concept, give examples, make lists, describe or define, reflect, justify a solution, write a summary, predict what might occur, or compare and contrast what they are learning. The simple fact is that if our students are not required to write, if they are not given feedback on their writing in the content areas, then they will not acquire the language and will not perform well on tests like the National Assessment of Educational Progress—considered to be the nation's report card.

Tests, quizzes, notes, and homework assignments should include writing. In mathematics, which is considered a language by many, vocabulary and notation are seldom tested at the upper grades, despite what the research suggests.

As part of the homework, teachers might ask students to write a procedure or formula or to explain a concept. The teachers might have the students write a procedure or formula two or three times to embed it in memory for multiple days. During note-taking, if something appears out of the ordinary—a "trick"—students should be given time to write an explanation of what occurred so that when they study their notes later, the problem makes sense.

Teachers might ask their students to use "concept cards." On one side of the card, the students write a basic concept or procedure, and on the other side they explain how they might address a variation in the concept or procedure.

By asking students to write about what they understood about a lesson or what caused them difficulty or confusion, teachers would gain insight into how they might address their own instruction to increase student achievement.

Writing in the content areas will cause students to think, reflect, remember, recall, organize, visualize, and analyze their thoughts. The components of writing are closely associated with the components of studying. Writing in the content area is different from teaching students to write. It is not intended to be graded based on grammar, spelling, and the like—although if students are careless in spelling or grammar, then the classroom teacher should take corrective action.

Writing in the content area helps teachers understand student thinking—it's another way to monitor student learning and understanding. If students are not using correct vocabulary or notation, then teachers should assist them by writing on the board important words or phrases that students should use in their writing. For example, when working with fractions, if students are using terms like "top number" and "bottom number," the teacher should require the students to use the words *numerator* and *denominator* instead. Feedback is important.

Instructional leaders should encourage their teachers to incorporate writing into their daily lessons by making suggestions and recommendations or by giving directions and providing helpful feedback on implementation of writing in the content areas. The writing process forces the students to reflect, think, and organize their thoughts.

Writing keeps students on task and reinforces concepts, procedures, vocabulary, and notation that teachers say they value. Teachers who regularly ask students to summarize the concept being taught have students who understand that math is more than just memorization. Reading, writing, and learning notation is extremely important to English-language learner students as well as to students coming from poverty. Language acquisition is a necessary component that leads to increased student achievement. However, that acquisition will not occur without repeated exposure—teachers orally using the language, students orally using the language, reading it, writing it, and speaking it.

THE STAR SYSTEM REVISITED

As we have discussed identifying items in the notes using the star system, it might be helpful for students if teachers identified questions on the homework using the star system as well. Place three stars by items that will be tested with no changes. Use two stars for problems that will be tested with minor changes—such as the specific numbers. Put one star on homework problems with greater variation, including modeling, some definitions, and procedures. Again, the idea behind the star system is to highlight what teachers expect students to know, recognize, and be able to do without narrowing the curriculum.

The homework and notes should reflect the instruction and help focus students on what they are expected to know, recognize, and be able to do. This focus helps students not only to organize their learning but also to study more effectively and efficiently.

POINTS TO REMEMBER

Knowing the importance of homework to increase student achievement, school administrators can easily monitor the types of assignments given to students at their schools. If they walk into a classroom and see a homework assignment on the board that contains only a page number and exercises, the supervisor can immediately improve the likelihood of student success by making recommendations that parallel my recommended assignments.

They might suggest or recommend that teachers create a homework assignment sheet for a week or a unit that includes reading, writing, definitions, explanations, pattern development, rules, procedures, and/or formulas, along with exercises, to help the students become more successful. These assignments also have the potential of taking the excuses off the table for students who don't do homework.

Changing the way we plan and assign homework will further enhance student study. It is not hard to do, and it is something we should do to help students better understand the concepts and skills being taught. Let's just do it!

6

TEST PREP, THE LIGHT AT THE END OF THE TUNNEL

PUTTING IT ALL TOGETHER

Good test preparation is almost nonexistent on teacher unit exams. Too many teachers review for a unit test the day before the actual test. There is a much better way to help struggling students succeed.

Test prep is where the rubber begins to hit the road. Students trying hard on test day, without the benefit of studying beforehand, will not enjoy the success that so many want to experience. So, while it is clearly the student's responsibility to prepare for a test in advance—to study—it is our responsibility to better help them prepare for the test and be successful. Just as coaches take responsibility for the preparation of their teams, teachers need to take responsibility for the preparation of their students.

QUIZZES

Classroom instruction is reflected in student notes and homework assignments, but another critical issue is further monitoring of student learning through the use of quizzes. The expectation is that quizzes should contain items related to vocabulary, notation, identifications, conceptual or pattern development, procedures, and exercises as well as problems and applications. Those are the very things that students would have seen and heard during instruction, written in their notes, and completed in their daily homework assignments. Quizzes emphasize what teachers expect students to know, recognize, and be able to do and allow for teachers to monitor student progress more formally along the way before the end of a unit.

UNIT TESTS

In a typical math chapter or unit, several strategies are introduced and taught. For instance, most algebra books will have units on factoring polynomials, solving equations, graphing, and so on that require decision making when all the varying methods are introduced. Students learn to solve systems of equations in a first-year algebra class by graphing, substitution, and linear combination. They learn how to solve quadratic equations by the zero product property, completing the square, and the quadratic formula. So, as students are quizzed on individual methods over a short period of time, many do well.

While students might clearly understand how to factor polynomials by each method as they are being taught to them, when grouped together they may all look alike to them, which often results in them not choosing the best method. This suggests that after teachers have taught the material in a chapter, they should take a day or two before a test to make sure students can mentally organize what was taught and know which method of factoring is most appropriate for each polynomial presented and why. Don't assume the students can do this on their own; most students cannot.

With quadratic equations, students again have to know not only the three methods of solving those equations but also when each method should be employed to make their work easier. Under what circumstances should the different strategies be used?

For example, look at these quadratic equations: $x^2 + 7x + 12 = 0$ and $x^2 + x + 12 = 0$. For most beginning students, they look very much alike. But the zero product property would be the most appropriate to solve the first equation, while to solve the second equation, the quadratic formula would be the best option. Those decisions can be made in seconds by looking at the coefficients, and students need to be able to make those decisions to be successful in math.

By teachers taking an extra day or two to explicitly teach students to differentiate between problems that look alike, students are more likely to see the big picture and should be able to make decisions that make math easier. Keep in mind our basic axiom: *The more math you know, the easier math is.* Choosing the right strategy for solving a problem makes the work much easier.

In addition to knowing how to attack similar problems, students need to know how those questions might be asked on different tests. *Find the answer*, *solve*, *find the solution set*, *find the zeroes*, and *find the roots* are all ways to say "Find all the values of the variable that make the open sentence true." Students need to know these mathematical synonyms.

PARALLEL-CONSTRUCTED PRACTICE TESTS

Students can learn more effectively when they know what they don't know—something a lot of people don't judge accurately.

Classroom tests, unit tests, are criterion-referenced tests. They are tests based on what is taught and learned in the class. As such, the contents of a teacher-made test should not be a secret. As part of the Components of an Effective Lesson, teachers are asked to not only state but also write the daily objective on the board—to be explicit so students know what they are learning. Testing should not be any different. Students should know exactly what is expected of them. After all, as teachers, many of us had to take exams. We prepared by studying older forms of the ACT, SAT, PPST, or PRAXIS. Not only did teachers study for these exams using practice tests, but chances are that those teachers wanted to know the rubric used in grading the tests.

Providing struggling students with parallel-constructed practice tests is another way of ensuring students know what to expect on their tests, and it also provides the teacher with another opportunity to monitor student learning.

Rather than scheduling a day for students to take a practice test in class, the recommendation might be to distribute a practice test to all students approximately halfway through the unit to be tested. Do not administer the practice test at that time. That practice test would be the test the teacher made up as part of preparing for the unit. Identify the questions students should be able to answer based on instruction and the questions that have yet to be covered. At the end of each successive class, be willing to answer questions from the practice test as well as identifying additional questions the students should be able to answer based on the new instruction.

Providing a practice test in advance, students can be more engaged in test preparation over a period of time that results in students having a clearer picture of what is expected. It also leads to increased student achievement.

Some might argue that providing a parallel-constructed practice test is teaching to the test. That's not true. The types of questions asked are based on the specification sheet and assessment blueprint. The items are based on what we expect students to know, recognize, and be able to do after instruction. Parallel-constructed practice tests don't have to give the exact test questions away, but the *types* of questions, the assessment blueprint, should be known by all—especially the students.

On a practice test, for example, students might be asked to add $\frac{5}{18}$ and $\frac{7}{24}$, a two-star problem; on the real test, different numbers would be used. On the practice test, students might be asked to write the procedure for adding or subtracting fractions, a one-star problem; on the real test, the procedure asked for might be for multiplying fractions. A specification-sheet question on a practice test might require the students to find volumes of solids. On the actual test, students could be asked to find the volume of a cylinder, rectangular prism, triangular prism, or pyramid. If that is called teaching to the test, that is good.

READY, SET, PRACTICE TEST

Just as I encourage teachers to provide students a copy of the practice test approximately halfway through the unit so they can clearly see the connection between the daily instruction and expectations, I recommend that teachers review the practice test, doing each problem, two or three days before the administration of the real test—not the day before the test. Students need time to reflect and process what they have learned and have an opportunity to ask for clarification, which cannot happen if the teacher reviews the practice test the day before the real test.

As teachers review for the test and pace the students through each problem on the practice test, they should also identify the questions on the practice test as one-, two-, or three-star questions. Ask students as they do each problem, "What could I ask if this question was a two-star? What could I ask for this question if it was a one-star?" As the students complete each problem on the practice test, the teacher should pace them and also do the problem to ensure all students are able to do the problem.

To build confidence, teachers must continually tell their students that they should expect to earn a 100 percent on the test because they know what is on the test and they are prepared. They are prepared because the instruction, notes, homework, and test preparation prepared them to demonstrate their knowledge. There are no surprises. Positive reinforcement, setting a high expectation, is important, and having them understand you believe in them will lead to student success.

Teachers should also continue to stress that it is not acceptable to miss three-star questions on the test. These are questions that have no manipulation or computation and that students need to be able to answer to do well on other test items. It should be made abundantly clear that any student missing a three-star question will be provided an opportunity to never, ever forget that definition, identification, formula, math translation, or procedure again. Expectations without consequences are meaningless.

The next day, based on their monitoring of the practice test, teachers should address any questions, concerns, or hesitancies they observed on specific questions as they walked the students through the practice test—probably giving the students more practice problems for those questions to further ensure their success. These additional days provide more time for students to fully process information, differentiate between seemingly similar problems, and provide teachers more time to clarify information based on their monitoring of student understanding from the day before.

It is very important that the practice test sets students up for success on the real test. That's where the rubber will hit the road. If time is not taken to fully prepare struggling students and they are not successful, then they will interpret all that encouragement and strategies as more of the same old stuff they have heard before.

To increase the likelihood this is occurring, school administrators should schedule a classroom observation during this test preparation. They should also consider dropping in the following day to confirm teachers are addressing areas in which students experienced difficulty. The big idea behind the "6+1" is to do everything possible to prepare students for the test, to be successful, and to see that translated into higher grades.

The following tests are examples of a practice test and a real test. Take your time and examine Test A, which we will use as a practice test. Then compare it to Test B, the real unit test. You will see they are constructed in parallel.

TEST B: Linear Equations and Inequalities

Name:_________________________ Date:_____________________

On Exercises 1 and 2 give an example of each property:

1. Associative property of multiplication:

2. Multiplication property of equality

3. Define: Distributive Property

4. List the order of operations:

5. Write out the following mathematical phrase in words: $\{x \mid x \in \Re \wedge x \leq 2\}$

6. Give the mathematical definition of absolute value: $|x| =$

7. Write the strategy for solving equations containing absolute value.

Find the following solution sets:

8. $5x - 4 = 36$

9. $7z + 4 = 39$

10. $\frac{c}{4} + 5 = 8$

11. $\frac{2y}{3} + 4 = 6$

12. $3(2x - 4) + 5 = 23$

13. $5t + 2 = t + 36$

Find the following solution sets:

14. $7(2x + 1) = 4x - 13$

15. $4(2w + 1) - 3 = 2w + 13$

16. $\frac{5c}{2} - \frac{c}{3} = 9$

17. $\frac{b+3}{4} - \frac{4b-5}{5} = -1$

18. $|x - 3| = 4$

19. $|2y - 1| + 4 = 13$

Solve and graph the solution:

20. $3x - 5 > 7$

21. $4(2z - 3) + 5 \leq 6z + 7$

22. Solve for L. $A = 2L + 2W$

23. Solve for b. $A = \frac{1}{2}(b + c)h$

24. Which number is *not* a solution of the inequality $2 - 3z \geq -4$

a. 0 b. –5 c. 4 d. 1

25. Fill in the following reasons using the properties of real numbers and properties of equality:

$2x - 3 = 17$	Given
$2x - 3 + 3 = 17 + 3$	____________
$2x + 0 = 20$	____________
$2x = 20$	____________
$x = 10$	____________

26. ***Write contact information for a parent or guardian. The information can be an address, phone number, cell phone, or email address. (CHP)

TEST B: Linear Equations and Inequalities

Name:________________________ Date:____________________

On Exercises 1 and 2 give an example of each property:

1. Associative property of multiplication:

2. Multiplication property of equality

3. Define: Distributive Property

4. List the order of operations:

5. Write out the following mathematical phrase in words: $\{x \mid x \in \Re \wedge x \leq 2\}$

6. Give the mathematical definition of absolute value: $|x| =$

7. Write the strategy for solving equations containing absolute value.

Find the following solution sets:

8. $5x - 4 = 36$

9. $7z + 4 = 39$

10. $\frac{c}{4} + 5 = 8$

11. $\frac{2y}{3} + 4 = 6$

12. $3(2x - 4) + 5 = 23$

13. $5t + 2 = t + 36$

Find the following solution sets:

14. $5(3x + 2) = 3x - 2$

15. $3(2y + 5) + 2 = 4y + 19$

16. $\frac{3c}{2} - \frac{c}{3} = 7$

17. $\frac{w+3}{4} - \frac{4w-5}{5} = 1$

18. $|x + 3| = 7$

19. $|2x - 1| + 7 = 18$

Solve and graph the solution:

20. $4x - 5 < 7$

21. $4(2p - 5) + 2 > 5p - 6$

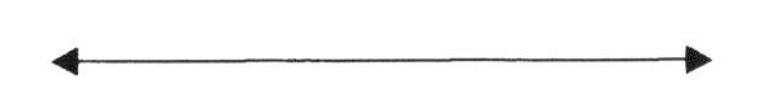

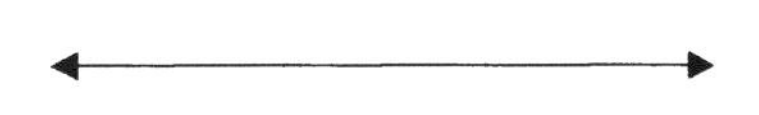

22. Solve for b. $P = 2a + 3b - c$

23. Solve for c. $A = \frac{1}{2}(b + c)h$

24. How many solutions does the equation $-2y + 3(4 - y) = 12 - 5y$ have?

a. None b. One c. Two d. More than two

25. Fill in the following reasons using the properties of real numbers and properties of equality:

$4x + 3 = 15$	Given
$4x + 3 - 3 = 15 - 3$	______________________
$4x + 0 = 12$	______________________
$4x = 12$	______________________
$x = 3$	______________________

26. Write contact information for a parent or guardian. The information can be an address, phone number, cell phone, or email address. (CHP)

Notice that the format is exactly the same on both tests. It's important to build confidence in struggling students that what you say you will test them on is what you do test. Also notice that, even though I have changed problems, the students can immediately see that problem 1 on the practice test, Test A, was asking about a property of real numbers, and on the real test, Test B, it asked a different question but one that was also about the properties of real numbers. Similarly, questions 20 and 21 on both forms of the test are about graphing. The actual problems are different, but they are of the same type. This is what is meant by parallel-constructed tests.

You will also observe that there are a couple of problems that did not change from Test A to Test B. Those would have been identified in the notes, homework, and practice tests as three-star (***) problems. Three-star problems are typically definitions, identifications, formulas, or procedures that students need to know to be successful on the test, but do not include computation or manipulation. The two-star problems would have minor changes, such as changes in the numbers used in solving the problem. The one-star problems are the ones that vary so the curriculum is not narrowed.

The last question on the test asks for parental/guardian contact information. It is a three-star question, and by providing that up-to-date information, students earn points. You might have spotted the letters CHP in parentheses. CHP is code for "Call Home Please." Struggling students typically do not like to have communication between home and school because it is generally nothing but bad news. The CHP coding allows students who feel like they were successful on the test to say please call home and let someone know I did well. Struggling students, students who have not enjoyed success in math, students who have always earned grades of D or F in math, really like the recognition of that success.

Teachers like recognition from their supervisors when they do well, and students do, too. Build on that success to create further successes.

TRANSPARENCY, CREDIBILITY, TRUST

Unit tests are criterion-referenced tests. Students should know exactly what is expected of them—there should be no surprises. By constructing the actual test using the same format as the practice test, with the same types of questions in the same order, students—especially those who struggle—will begin to believe they can be successful in mathematics. They will see their classroom teachers as people who want them to succeed, people they can trust.

Student confidence, morale, and performance will suffer if students lack confidence in their teachers. The solution is clear: Teachers must demonstrate transparency in what is taught and tested. Transparency leads to credibility, which in turn leads to trust. There should be no secrets. Fairness counts.

Practice tests and unit tests should have the same format. Students should easily be able to determine what will be expected of them on their unit test by looking at the practice test and perhaps using the star system.

Making the connections among classroom instruction, student note-taking, homework, quizzes and practice tests, and tests is a critical and practical thing teachers can do to help students learn—to help organize their learning. These are absolutely necessary for the success of struggling students, to help them be more effective and efficient in their study of mathematics.

The use of practice tests has four main functions:

1. Creating a practice test before instruction begins suggests that teachers know what to expect their students to know, recognize, and be able to do before instruction begins and that they are prepared for the unit
2. It explicitly informs the students what they need to do and acts as a blueprint for success
3. It can act as a motivator for improved performance in the classroom
4. It provides teachers with another opportunity to monitor student learning and address deficiencies to increase student performance and achievement.

As students gain confidence, are better able to determine what to study and how to study for exams during the school year, and demonstrate success on unit tests, teachers might ask more one-star questions than three-star and slowly begin to construct practice tests and real tests that are not constructed completely in parallel.

We already prepare our students for tests. Using parallel-constructed tests to prepare struggling students for unit tests will increase student confidence.

Using a parallel-constructed practice test that is reviewed a few days before the actual test summarizes the instruction and notes to three or four pages and acts as a light at the end of a tunnel that can serve to motivate struggling students.

POINTS TO REMEMBER

School administrators should ask teachers for a copy of their practice tests before instruction begins not only to ensure the teachers are prepared but also to determine if the tests measure the academic standards; are reflective of questions on tests such as the ACT, SAT, NAEP, and so on; and have questions worded with the same formality students will encounter on high-stakes tests. By checking these practice tests before instruction begins, the school administrator can resolve issues before they become problems involving student grades.

If teachers are not giving a common unit exam, the school administrator should compare and contrast the individual teachers' tests to make certain the tests are fair and the grades will be portable because they have the same difficulty level.

Creating practice tests and reviewing them a few days before the actual unit tests results in increased classroom performance and higher grades. It builds students' confidence and provides struggling students a blueprint to be successful. School administrators should also make an effort to be in classrooms during test prep to verify everyone is adhering to the expectations.

Following the test preparation recommendations will result in better test scores and grades. It's good for students to experience success, so let's teach how to prepare to be successful. Let's do it!

7

THE TEST, WHERE THE RUBBER HITS THE ROAD

SETTING A DATE FOR A TEST

Tests should reflect what we expect students to know, recognize, and be able to do. Test results are a reflection of instruction. They can either move a class forward pursuing excellence or set a class back. Teachers' beliefs followed by their actions will often determine if the foot is on the accelerator or brake.

As teachers do their long-range planning or use the guide for preparing for a unit described in chapter 2 to prepare instruction, benchmarks are identified. Setting benchmarks is important, so that teachers can plan to teach their assigned curriculum to mastery. As part of that planning, testing dates are also typically identified. Since teachers are in the planning stages, the dates should be flexible. If a test is scheduled for a specific Friday and the teacher determines on the preceding Tuesday that the students are not ready, the test should be postponed. There is no sense administering a test to students who are not prepared for it. Remember, build success on success.

Based on student work, guided and independent practice, class participation, homework assignments, questions students are asking, quizzes, and so forth, teachers should be able to accurately predict student performance. That insight should be used when determining if additional time and instruction will benefit the students in preparing for the test.

Postponing a test requires the teacher to move forward with new instruction, while giving the students a few more days to successfully prepare for the test. Teachers incorporating my recommendations could use the last five to seven minutes of each class to address the deficiencies that led to postponing the test. It does not mean to stop, reteach the entire lesson, and then give the test. Teachers that stopped all new instruction would run the risk of not adequately covering the curriculum assigned to them—creating more student deficiencies that someone else will have to address or will eventually result in adult-caused student failure.

Teachers would be more inclined to give students additional time, when warranted, if they viewed the test results as a reflection of their own instruction. Teachers who do not monitor student learning as they are teaching might not know that their students are not understanding the material being taught. A teacher that is surprised by test results probably needs to pay closer attention to questions being asked in class or to questions that are *not* being asked, reading body language and facial expressions, performance on homework and quizzes, guided practice assignments, and the practice test, as well as the discussions taking place in class.

TEST TEMPLATE

Students know what's important in a math class by what is tested. If math tests are nothing more than doing exercises, these students won't value the language; won't learn the vocabulary, notation, and procedures; and will continue to struggle.

Tests drive instruction. In math, I have recommended that there be a balance in the delivery of instruction that is matched in the assessment of student learning. Balance has been defined as:

- Vocabulary and notation
- Concept development and linkage
- Memorization of facts and procedures
- Appropriate use of technology
- Problem solving

To transform struggling students into successful students, they must understand the importance of knowing the language of math.

Students should earn credit for information they know. The testing template recommended for working with struggling students is to have the first five or six questions—20 to 30 percent of the test—be based on definitions, listings, formulas, properties, procedures, identifications, and translations. An example can be seen on the tests shown in chapter 6. These are typically three-star questions. This part of the test does not contain any type of computation or manipulation. It is very straightforward, based on what was taught and communicated to the students they were expected to know.

Educational research indicates that students who cannot verbalize information generally don't "own" it, which translates to not knowing it. That contributes to students not being able to do the problems on the test—resulting in failure. If a question on the test is to write the "distance formula" and the student did not memorize it, the probability of that student finding the distance between two points on subsequent questions on the test are pretty low. In contrast, if questions asked at the beginning of the test include definitions, procedures, formulas, and so on, the probability of students who correctly answer these questions being able to correctly answer subsequent questions increases. If teachers adopt such a template, a second benefit is that students earn credit for the information they know.

The second part of the template contains computation, manipulation, or solving equations that are related to the first part of the template based on the specification sheet and academic standards. Part 3 of the test is the conceptual development, linkage, and understanding part of the test. Part 4 is generally the problem-solving portion of the test.

This template helps students become more successful and creates the realization that vocabulary, notation, and being able to read, write, and speak the language of math helps them better understand the mathematics they are learning.

Parts 1 and 2 of the template are almost always there. Parts 3 and 4 are used when they contribute to the goals and objectives of the unit based on the standards.

Students are always told to check their answers in a math class. To be consistent with that advice, tests should generally have twenty to twenty-five questions so the students have time to do the test and to check their answers.

If there are items on the test that have a great tendency to trip students up, placing a caution mark beside the problem on the practice test to encourage them to pause and think will help struggling students be more successful. An example might be a problem that contains the expression -5^2, which is -25; many students have a tendency to confuse that with $(-5)^2$, which is $+25$.

While tests are used for grades, they should also be used to monitor student learning and the effectiveness of instruction. When students are not succeeding, a change in instructional strategies or resources may be in order.

A test design should foster success and encourage studying. Using the test template, the first five questions on the following test are regurgitation type problems. There is no computation or manipulation required—just memorization. Hopefully there was understanding with that memorization.

The accompanying practice test follows that template. Using that template, it is easy to see how the regurgitation-type questions—definitions, formulas, algorithms, and so forth, information students need to know to be successful—also address other questions on the test. Usually, those questions are in the three-star category.

Question 2 is a one-star problem, not three-star, because we don't want to restrict or narrow the curriculum and want students to know how to solve quadratic equations. There would typically be three methods of solving quadratics equations taught: completing the square, using the zero product property, or using the quadratic formula. Since we want students to know all three methods, we will let students know that on question 2 they will be asked for the algorithm or formula of one of those methods. Again, because it is in the beginning of the test, using the template, these are questions students need to know that don't require computation or manipulation.

Questions 3 and 4 ask for a formula and other information students need to know. By memorizing the formula and other pertinent information, the students receive credit for what they know. Knowing that also sets them up to be able to answer other questions on the test. In other words, if they could answer question 3, then the probability of answering question 6 correctly is increased. If they could not answer question 3, then it would be very unlikely they could answer question 6, and you would know why they got it wrong. Knowing how to answer question 4 correctly sets students up to be successful on subsequent questions as well.

Some teachers might argue that the regurgitation questions are embedded in the computation/manipulation questions on the test. But the template allows students to earn credit for what they know; having those questions

Algebra II, Chapter 5 Practice Test

Name______________________________ Date____________________

Period 6

WRITE YOUR FORMULAS & SHOW YOUR WORK

1. ***Write the General Form of a Quadratic Equation.

2. *Write the Quadratic Formula.

3. ***Write the formula for the discriminant and explain how it is used to determine the number of roots.

4. ***Write the formula for finding the axis of symmetry and vertex (x-coordinate) in a quadratic equation.

5. ***$i^2 =$

6. **Use the discriminant to determine the number and types of roots in the equation $y = 2x^2 - 3x + 7$.

7. **Find the vertex of $y = x^2 - 10x - 13$.

8. **In the equation $y = 4(x + 6)^2 - 4$, identify the vertex.

9. **Write $y = x^2 - 6x - 4$ in vertex form (hint: complete the square).

10. **Graph $y = x^2 + 2x - 3$ using the vertex and symmetry around the axis of symmetry.

11. **Graph $y = x^2 + 6x - 4$ using the vertex and x-intercepts.

12. *Solve $2x^2 + 15 = 13x$ by the Zero Product Property.

13. **Solve $x^2 - 3x - 10 \geq 0$ and graph the solution set using the Zero Product Property.

14. **Graph $y < x^2 + 6x + 7$.

15. **Graph $y \leq -x^2 + 8x - 10$.

16. *Write in Standard Form: $(3 - i) + (1 + 5i)$

17. *Write in Standard Form: $(4 + 2i)(3 - 5i)$

18. *Write in Standard Form: $(2 + 3i) / (5 + 2i)$

19. *A real estate developer estimates that the monthly profit p in dollars from a building s stories high is given by $p = -2s^2 + 88s$. What height building would he consider most profitable?

20. ***Write a home phone, cell number, email, or home address to contact your parent or guardian. (CHP)

embedded in a problem does not. Also, having those questions on the test provides teachers a greater opportunity to understand student deficiencies.

ASSESSMENT DRIVES INSTRUCTION

School administrators should collect data on the grade distribution of students in classes after the first teacher-made unit test is given in September. If the percentage of students earning a grade of D or F is high, then that is clearly an early warning sign that a teacher may need assistance. It is clear the students need help. Administrators should not be waiting for the completion of the first quarter or semester grades before assessing the needs of students and teachers. By then, credits required for graduation may be in jeopardy.

The recommendations outlined in this book should assist teachers of struggling students. After observations, school administrators could provide feedback on making instruction more understandable and make recommendations with respect to student note-taking, homework assignments, and creating parallel-constructed practice tests. In other words, the administrator should make sure teachers are helping students organize and focus their learning to help them study more effectively and efficiently. That will result in increased student performance.

Monitoring student progress frequently and systematically helps teachers identify strengths and weaknesses in student learning as well as in instruction. Assessing student work comes in many forms, but teachers need to know the answer to this question to improve their instruction and address student needs: "What do my students know, and how do I know they know it?"

There is often a disconnect between what teachers say they value in mathematics and what they test. When talking with classroom teachers, they will indicate the importance of vocabulary and notation to be successful in math. But when you look at their own classroom tests, typically there are no questions on vocabulary or notation. Teachers will tell you how important it is to have student understanding, but again, when you look at their unit tests, there won't be questions that require open-ended answers that measure student understanding of the concepts being taught.

In fact, many teachers will readily admit that students have raised their hands in the middle of the development or explanation of an important concept only to ask, "What's the homework assignment?" Now, why would a student interrupt instruction to ask such a question? The answer is that the students know what the teachers value better than the teachers themselves do. Students know teachers value what's being graded—that is, what's on the homework, quiz, or test. If teachers don't ask questions dealing with conceptual development, linkage, vocabulary, and notation on tests, then chances are students won't spend time studying it.

Teacher-made tests should reflect what is taught and valued in mathematics education. For example, while many teachers say mathematics is a language, this may not be reflected on their tests. If we value students' ability to verbalize their knowledge, then definitions, identifications, and procedures should be part of tests.

Many of the rules in math don't make a lot of sense standing alone. For example, when students add fractions, it seems natural to add both the numerators and denominators, and yet they are told not to add the denominators. If students learned that a fraction is a part of a unit, made up of a numerator and denominator and that the denominators told them how many equal parts makes one whole unit, it would make sense not to add the denominators so they know how many equal pieces makes one unit.

As another example, while it is important for students to know they cannot divide by zero, they should be able to give some rationale behind the rule, besides saying, "That's what my teacher said." Having the students write a brief explanation using the definition of division will clarify their understanding. Manipulation of data, vocabulary and notation, open-ended questions, problem solving, and appropriate use of technology should be included on tests. Also, to encourage students to review and reinforce previously learned material, teachers should make their tests cumulative.

Teachers using the specification sheet, assessment blueprint, and benchmarks discussed in the earlier chapters would be more likely to have balanced assessments that measure what they say they value in math education. Teachers should not expect of their students what they are not willing to inspect.

Good teachers prepare their students to succeed. In preparing students for tests, teachers should provide tips on how to study. For instance, students sometimes confuse newly introduced terms such as *domain* and *range*. It might be helpful if teachers wrote an ordered pair (x, y) and pointed out that those are in alphabetical order as are the pairs (domain, range), (abscissa, ordinate), and (horizontal axis, vertical axis). Those connections might help students remember.

Teachers should also take the time to help students differentiate between problems that look alike. For example, while students might learn several different methods of factoring polynomials in algebra, they may not be able to determine an appropriate method of factoring when a mixture of factoring problems is presented. Students have to be taught how to recognize differences and when to apply each method. Comparing and contrasting leads to increased student achievement.

All too often, teachers successfully teach each method to their students on a section-by-section basis, but don't take the time to teach them to compare and contrast these problems so they know which method to use. When the students perform poorly on the test, the conclusion reached by many teachers is that the students did not learn how to factor. But the reality is that the students learned what the teacher taught—to factor—but did not learn what they were *not* taught—to differentiate between problems that looked similar and to factor using the appropriate method.

Tests are formalized vehicles to not only evaluate student learning but also act as an assessment tool. As such, tests provide students a blueprint to increase their knowledge. Teachers should use assessment information, particularly questions answered incorrectly, as one way of increasing student performance. Addressing these deficiencies will result in increased student achievement, as will addressing their own instructional practices.

ASSESSING STUDENT WORK

In order to address student deficiencies, teachers need to know what students know. Earlier, I mentioned questions that students should not be allowed to miss on a unit test—three-star questions. The fact is, some students will choose to miss them. But missing them is not acceptable, and there needs to be a sanction that says just that.

Students who miss a three-star question—a question that was identified in the instruction, notes, homework, and test preparation, a question without any computation or manipulation, strictly memorization—need an experience that will not soon be forgotten. Writing the question and answer a significant number of times seems to do the trick for almost all students—enough times that they will always remember your name and the item they missed and be able to tell that story to their great-grandchildren. Education is important, so it should be important enough to have consequences for missing a three-star question.

Since the teachers have already called and informed the parents of how they will help their kids not only succeed in math, but excel, it's always wise to call the parents ahead of time and remind them of their support and how this activity (sanction) will help their kids in the long run. The reason a student misses a three-star question is because they just plain did not study.

On many state-mandated tests, it would be pretty difficult, if not impossible, to determine deficiencies based on the makeup of the test, so unit tests are important. Often, one test question is meant to measure multiple state standards, including a student's ability to compute. If test items are all made up of word problems, how can a teacher determine if a student missed a specific question because the student

- does not speak English,
- has a reading comprehension problem,
- did not understand how to solve the problem,
- did not know how to compute using that number set, or
- made a simple computational mistake?

Without knowing what students know, teachers will frustrate themselves and their students by having to reteach things the students already know. Boredom often results in students getting off task resulting in classroom management problems. That reteaching also takes time—time teachers don't have.

Identifying deficiencies and addressing them will provide teachers with more time to teach the curriculum assigned to them to mastery. In the factoring example given above, if the teachers retaught the five methods of factoring taught in a first-year algebra class, they would spend a lot more time than if they identified the deficiency as being able to distinguish between polynomials.

Teachers complain all the time about their students not knowing their basic arithmetic facts. Yet if teachers looked at each operation, they would find that students actually know most of the facts. For instance, if you asked teachers if their students can multiply by 1, 2, 3, 5, 10, and 9 and can multiply doubles, the answer to each of those questions is generally yes. That means the students know most of their facts. This information allows teachers to concentrate on student deficiencies. Knowing the actual deficiencies allows teachers to spend their time reviewing and reinforcing what they don't know—in this case, multiplying by 4, 6, and 8. And not even all those need to be addressed, because the students, using the commutative property, can multiply those numbers by 1, 2, 3, 4, 5, 9, and 10. That means the teacher needs to concentrate on 4×6, 4×7, 4×8, 6×7, and 6×8 as the facts that need to be addressed. By identifying what the kids don't know, the teacher is able to cut down quite a bit of unnecessary reteaching. That knowledge of what students know takes a lot of frustration out of teaching and does not bore the students to death.

Another example of assessing what students know might be adding fractions. On a test, if a student could successfully add $\frac{1}{4}$ and $\frac{1}{3}$ but could not combine $\frac{5}{18}$ and $\frac{7}{24}$, some teachers might deduce that the student does not know how to add fractions. Another teacher, upon closer examination, might conclude that the student has the procedural knowledge, because he or she successfully added the fractions on the first problem, but is getting the second problem wrong because of difficulty finding a common denominator.

Rather than reteaching the procedure for adding fractions, good teachers will know what their students know and concentrate on finding common denominators. Teachers find out what students already know and what they still need to learn by monitoring and assessing student work—not by mere perception.

Introducing new concepts by linking them to previously learned concepts and outside experiences has been emphasized for a number of reasons. One of those is familiarity with language, making the students more comfortable in the concept being introduced. The importance of the formality of language has also been discussed. On teacher-made tests, teachers must be on guard to use on their own classroom tests the more formal language that students will find on standardized tests. Otherwise, students might not recognize the question being asked as one that was taught in school.

For example, on an algebra test, a teacher might ask students to "solve" an equation. On a college entrance test, students would be asked to solve an equation by "finding the solution set." This difference in language might result in a disconnect in what was taught and learned in class and what is being asked on a test. Students not recognizing that these different directions mean the same thing might miss a problem they really know how to do. Unfortunately, this kind of mistake might result in people in the community believing students are not being taught these concepts and skills in school.

If teachers truly understand the importance of vocabulary and notation and its relationship to increasing student achievement, then they would ask students to evaluate $f(2)$ if $f(x) = 5x - 3$. Students should have been taught and tested on the math translation. "The value of f at x is $5x - 3$." Or "the value of f of x is $5x - 3$." To evaluate $f(2)$ means to "find the value of f when $x = 2$."

If teachers tested what was taught, the tests could be used to explicitly identify deficiencies. For instance, students asked to solve counting problems should be encouraged to use a calculator. A correct answer would suggest the student knew how to solve the word problem. Students not using a calculator who answered incorrectly may not have understood the problem, not have known a particular formula, not have known how to use the formula, or have made a simple arithmetic mistake. If teaching students to solve counting problems is important, then test the students on that and allow them to use a calculator. If it is important for students to know a formula such as $5^{C}3$ (I think it is), then have them write the formula and evaluate it. And make sure they know how to say it—"a combination of five things being taken three at a time."

Experienced teachers can predict common errors students will make. For example, all algebra students are taught to solve quadratic equations by the quadratic formula. Most students will memorize the formula, identify the values of a, b, and c, plug them into the formula correctly, and evaluate the resulting algebraic expression, then simplify the last step incorrectly. The most common mistake students make when using the quadratic formula is simplifying the fraction without factoring first. When this occurs, some teachers might conclude that their students did not learn what they were taught. The fact is, however, if they memorized the formula, could identify the values for a, b, and c, were able to plug them into the formula correctly, and successfully evaluated the expression, but made a reducing error, then the students did learn what was taught. The teacher needs to address reducing or modify their instruction so students are more successful; for instance, the quadratic formula could be written as a sum of two fractions.

POINTS TO REMEMBER

To determine whether teacher-made tests are fair, balanced, and consistent and cover the curriculum and whether the grades earned in different classes are portable, site administrators need to compare the content and achievement on grade-level or unit tests given by different teachers testing the same topics. This should be a common practice—a practice that almost always necessitates a follow-up discussion by the teachers and administrator.

The administrator should check to see if the actual test was constructed parallel to the real test. If the site administrator did not see the balance in the assessment to match the balance on the specification sheet, consistency should be a point of discussion with the teacher. If one teacher asked students to reduce a fraction like $\frac{4}{6}$ and another asked students to reduce $\frac{111}{123}$, a supervisor might ask about the fairness of these questions. Administrators should also ask teachers questions about the test items on teacher-made tests to ensure those questions reflect the types of questions students will see on high-stakes tests such as the semester exams, college entrance tests, and so on.

If students of specific teachers are experiencing a high rate of grades of D or F, then the school administrator should provide assistance in the form of recommending instructional and assessment strategies to improve instruction. That improved instruction should also be reflected in increased student performance on tests.

Creating a test template that gives students credit for what they know will increase student grades. If that template also helps students address other problems on the exam, then that will also increase student performance. Treat students the same way you would like other teachers to treat your children: build success on success. Let's make this work for all of our kids. Let's just do it.

8

RELATIONSHIPS AFFECT PERFORMANCE

One of the greatest concerns expressed by policy makers has to do with closing the achievement gap while improving student classroom performance and achievement. To accomplish this, as we have discussed, a number of elements have to be addressed. One that seems to get lost in the fray is the importance of building positive student–teacher relationships.

INFORMED PARENTS

One good way for teachers to begin the school year is by contacting the parents before anything could possibly go wrong to introduce themselves and explain what *they* as teachers are going to do to help the child become successful. The phone call is not about what the parents will do, it's about what teachers control—that is their actions.

Parents want their kids to succeed in school. The problem is that they get over their head in their other duties as a parent or at work. So educators need to understand that parents, grandparents, and guardians have other duties that interrupt their good intentions with respect to schooling.

What is needed for students to be successful is informed and cooperative parents that believe that we will do everything we can to help their child succeed in school. To accomplish this, making a phone call home is a good first step. Introduce yourself informally by name, as their child's math teacher.

Before continuing, ask if this is a good time to speak with them for ten to twelve minutes, and if it is, follow the outline below.

- Introduction
- Pleasure teaching your son/daughter, nice young man/lady
- Explanation, how I intend to help your child succeed—Instruction
  - Clear instruction, linkage, memory aids
  - Notes, star (*) system, very prescriptive
  - Homework, comes from notes and instruction
  - Oral recitation, procedures, and formulas
  - Practice tests, star system
  - Study/flash cards
  - Reviews
- Permission to use those strategies/consequences
- Parental help
  - Know when tests are scheduled
  - Examine student notebooks
  - Use flash cards to help study

Quickly explain how much you enjoy teaching, how long you have taught, and what a pleasure it is to have a nice young man or lady like theirs in your class. Also explain how you want their child not only to succeed but to excel in math. And with their understanding and cooperation, we can make math their favorite subject.

Ask if their kids have ever had trouble in math. If they have, which is generally the case, explain what you will do to make sure instruction is more understandable by using concept development and linkage to help the students have increased knowledge and better understanding and comfort with the math they are learning and how it is applied.

Ask if the kids have ever complained about not being able to do their homework exercises. When the parents answer in the affirmative, discuss how you will do your best *never* to send students home with an assignment they cannot do. Go over the homework format, explaining that the first four or five questions on any given homework assignment can be answered by just referring to the notes for definitions, identifications, procedures, and explanations. Also explain how if the child can answer those questions, the probability of being able to do the exercises correctly increases dramatically.

Explain how the star system in the notes, homework, practice tests, and assessments sets their kids up for success. After a brief discussion of this, parents really begin to believe you are a good person who wants their child to succeed.

This phone call is about building a relationship with the parent, building confidence and trust in you as the classroom teacher. It lets them know you want their child to experience success as much as they do and that you will be caring and fair, but also demanding.

After gaining this belief and trust, ask for permissions and consequences. That is, ask the parents if would it be okay, if their child refuses to participate in class-wide oral recitation to embed important formulas, definitions, or procedures in short-term memory, the child is made to write that information twenty to twenty-five times as a consequence.

Explain that the three-star questions are in the notes, homework, practice test, and real test—unchanged—that the tests are parallel constructed, and that knowing these items marked with three stars is very important for student success in the class. As such, if their child misses a three-star question on a test, you will ask the student, with the parents' support and approval, to write that information so that when they are eighty-four years old talking to their great-grandchildren, they will still remember that information. My experience with this is that most parents laugh, ask me again about three-star questions—making sure their child is not going to be treated unfairly—and then agree.

The phone conversation suggests to parents that you care enough to call. It provides classroom teachers with an opportunity to come across as caring educators who will do all they can to help their students succeed. It also sends a message to students—a message that indicates that if they get out of line in class, you will surely call home.

By the way, kids don't like the idea that you call. One reason is that the parents talk to their kids about how fair you are going to be, how they like you and believe in you. It is going to be a hard sell for students to tell their parents they aren't doing well because a teacher does not like them.

The bad news is that my experience last year was the same as my experience twenty years ago. You only get through to 40 percent of the parents on the first attempt. So calling home is time consuming: ten to twelve minutes per completed call, plus all the attempts.

Parents should be informed that they can be more involved in their student's work by physically checking and using the child's notebook every night. They could ask their kids to recite definitions, algorithms, and formulas, or they could ask them to do a problem that was written in the notebook as an example. Parents could ask them to explain the "why" behind the problems. Parents should also be clear that all the homework assignments will be written in the notebook as well the scheduled date for the next unit test. Let the parents know what they can do to stay positively involved in their kids' education.

That early connection often pays huge dividends on three fronts: classroom management, classroom performance, and increased achievement for students. Schools and school districts cannot afford to pay for such a positive public relations campaign; imagine teachers calling home to say we can help your child succeed.

STUDENT SURVEYS ARE TELLING

Teachers and math departments would benefit by employing student surveys about the instruction they are receiving. Surveys would further involve students and build on relationships by giving students some input in teacher evaluations. Student surveys are very revealing about teacher performance that some studies indicate accurately reflect student achievement. Students know whether teachers care, not by what they say but what they do.

Students know if there are classroom management issues because they are there every day to observe what occurs. Students know that some teachers explain concepts and skills and make learning easier. They know whether teachers can give good reasons for learning material, to create interest and enthusiasm in a subject. Students can tell you if the work is too easy or challenging, and they can also tell you if teachers take the time to make sure students understand what they are learning. Not involving students in the feedback process can result in school administrators missing practices that could lead to increased student achievement.

Having students be part of the process suggests we care and are serious enough about their education to include their input to make changes in our practices that will benefit them. It also leads to building stronger student-teacher relationships.

LAW OF RECIPROCITY

Many teachers have had the experience of having a student not like them and decide not to work, flunking to teach their teacher a lesson. The fact is, research suggests that students will work for teachers for no other reason than loyalty. As the professional, educators need to take advantage of that knowledge and talk to their kids. Teachers also need to watch how they talk to them. They need to be positive. Rather than saying things like, "If you don't do your homework, you will fail," they need to say, "If you do your homework, you will be successful."

Remember, treat your students the same way you want your own sons and daughters treated by another teacher. Talk to the kids. While you are not their friend, you can be friendly and can act as an advisor. Talk to them about sports, their social life, the dance, game, or weekend. Form a bond that suggests to the students that if they stopped coming to school, someone would miss them—that you care about them.

Remember this law of reciprocity: *People you like generally like you; people you don't like generally don't like you either.* If you have ever been in a long-term relationship, your partner may have expressed frustration about you not expressing your feelings in words. They want you to say you love them even though you think it should be just understood. The same is true in your classroom: You need to tell your students you like them, that you want them to be successful.

Reading my evaluations while teaching at the university, many of my students would fill out their anonymous evaluations and comment about how much I liked them, how I liked teaching, and how I wanted them to succeed. A colleague once asked me why my students felt that way about me. The answer was simple: I did my best to tell my students at least twice during the semester that I liked them, I liked teaching, and I wanted them to succeed.

CLEAR EXPECTATIONS

Teachers and administrators must explain expectations explicitly and give examples. Don't have a twelve- to fifteen-year-old interpreting those expectations. If you want the students to be in class on time, does that mean running in the door as the bell rings, in their seats, or in the seat with book or notebook open and ready for instruction? You need to tell them, or you will be frustrated the rest of the school year.

Build trust with your students. Make sure they know you are there for them. For instance, grading papers is not about taking points away from students. It should be about finding out how much they learned and helping them become more successful. Don't get caught up in arguing about points deducted in a test. If a student deserves the points, give them.

And while we are talking about testing, teachers should make testing as much a reflection of their own instruction as student preparation. If students are failing, the first place a teacher should look to find the problem is in the mirror . Remarkably, there is a relationship between what students learn and not only what they are taught but also how they are taught.

TEACHER EXPECTATIONS IMPACT STUDENT LEARNING

Many students sit in the back of the room because they want to be left alone. These unsuccessful learners could be classified, by their behavior, as reluctant learners. Teachers know them and are often thankful when those students sit in the back—quietly.

And while teachers do feel thankful they are not being disruptive, we need to refer to the "My Kid" standard and put ourselves in their parents' position. When children are four and five years old, they see their parents as the smartest and strongest people in the world. By the time these same kids reach sixth grade, they begin to view their parents as not very hip, out of date, or sometimes just not smart. As those youngsters reach twenty-five, it seems the parents start to regain some of the intelligence their offspring thought they lost when they were in their teens.

Students in middle school and high school tend to rebel against or question authority, which more often than not includes their parents. They would rather listen to the advice of friends than their own parents. Teachers need to help these parents with their children just the way they would like some other teacher guide their own kids.

Would a classroom teacher want his or her child to sit in the back of another teacher's classroom napping during instruction? If they would not want that for their own child, using the My Kid standard, they should not let that happen for someone else's children either.

Student–teacher relationships are important. Teachers need to earn the trust, confidence, and respect of their students by constantly communicating with them, encouraging them to be successful, and showing them how they can be successful.

Students who sit in the back of a room napping or seemingly ignoring instruction appear to be defiant. However, upon closer examination, teachers might just find out that that behavior is a defense mechanism, one that allows them to escape embarrassment or ridicule—ridicule for not being successful, for being seen as stupid.

Many of these students have not experienced success in the math classroom and have lost hope they ever will succeed in mathematics. Without teacher intervention, these students will be doomed, and underachievement in math may become a way of life. Too many students equate success in mathematics with being smart, not with hard work. That is sometimes unconsciously reinforced by classroom teachers who recognize students who learn quickly as having higher ability than those students who have to work harder or longer to complete a task or assignment. Speed at completing tasks appears to have replaced effort as a sign of ability.

High levels of effort might even carry the stigma of low ability—not being seen as smart by their peers—which results in students not expending the effort needed to achieve to their full potential. Once students begin believing they have failed or are failing because they do not have the ability, they lose hope for future success and stop trying. Hard work and effort have to be recognized by teachers as critical to student learning.

Teacher expectations can often be interpreted by listening to their own commentary. When teachers talk about "those" students, not "my" students, there appears to be a disconnect between what they do in the classroom and what students learn. You hear these teachers lamenting that these students are absent often, won't come to class on time, won't do homework, won't take notes, and so on. What's interesting about those commentaries is that some of those same students can be observed going to *other* teachers' classes on time, doing their homework, and taking notes. Knowing that these students will work for other teachers suggests that teachers do make a difference and that teacher expectations and their relationships with students have an effect on what students are willing to do in a classroom. If that is not true, then we should just consider going to video lessons.

STRUGGLING LEARNERS

Student expectations impact learning. People enjoy participating in activities when they are able to participate successfully. Some of us who may be monotonic may be reluctant to sing with a group of friends because we don't feel like our voice is that good. Some people really enjoy a game of chess or putting puzzles together, typically because they have experienced a certain amount of success or feel that, with additional time, they can be successful. The bottom line is, most people will spend time on activities in which they have experienced success or have an expectation of success.

The same can be true of students. They shy away from activities in which they have not been successful and feel like they will not succeed. Students who have not experienced success in mathematics might be viewed as reluctant learners. They are typically forced to enroll in a class—a class in a subject in which they have not been successful—and are expected to perform with a certain amount of interest and enthusiasm. Their reluctance to participate is the same reluctance adults have when asked to sing, speak before a group, or otherwise participate in an activity they don't enjoy or don't visualize themselves as being successful.

Educators need to better understand the importance of building success on success. Students who experience success in math will generally spend more time on math than students who cannot see the light at the end of the tunnel.

Classroom teachers must make an effort to build a positive relationship with all students—especially with unsuccessful learners—to build confidence and trust so the students won't feel threatened if they try. They must try to build a relationship by communicating with those students daily, resulting in those students feeling they can be successful and would actually be missed if they did not come to school.

To develop such a relationship, teachers need to talk to their students outside the classroom, in the halls and the lunchroom, at ball games, dances, and the store. Since many of these students have confidence problems, teachers might have to schedule a one-on-one conference with individual students during the school day—during another teacher's class, before or after school.

Teachers would be wise to have the conference with these students who are disengaging in a nonthreatening atmosphere. For example, teachers might sit with the student at a table, not behind a desk. Teachers should discuss what they are able do to help the student succeed in his or her class. In other words, the teacher must be part of the plan, too. They also need to elicit from the students what they are currently doing and offer constructive suggestions on how to more efficiently and effectively use their time. Once it is determined how students are using their time, the teacher should encourage them to do more. The teachers need to think about what they would like another teacher to say to motivate their own child and make the same kinds of comments in a friendly, helpful manner.

Classroom teachers must build success on success. To do that, they must teach students how to be successful. Here are some suggestions to help students succeed in mathematics classroom.

1. Neatly copy in your notebook any problems that are put on board. Be sure you understand each step of the problem as it is being explained. Ask questions to clarify any step that you do not understand. Do not wait to have the point explained at a later time.
2. Always try to do as much of the assignment as possible without help. To a great extent, the amount you learn is dependent on how well you have worked independently. When you practice a skill, it is more likely to become part of your long-term memory. Relying excessively on the teacher, or anyone else, to answer questions and to solve all the problems could result in a lack of understanding. If you are still confused after making your best effort, consider discussing the problem with a classmate.
3. It is necessary to spend time studying at home in order to reinforce what you have learned in class. Don't think that once you have obtained all the answers on an assignment, you are through with the

material. After completing an assignment, review the concepts with the idea that you will be expected to know the material on a test. By studying at home, you will discover what you do not understand and will be ready to ask questions in class the next day. Students who have done little studying on their own frequently know so little that they are embarrassed to reveal their ignorance. They are often afraid to ask a question that they feel everyone else in class can already answer.

4. When material you have already learned is being discussed, use the opportunity for "overlearning." Try to work a step ahead of the person presenting the problem.
5. DO NOT WASTE TIME IN CLASS! Most of your learning will occur during class time; it is foolish to waste this time. Each class period gives you an opportunity to concentrate on learning a specific concept, to correct your mistakes, and to direct your learning efforts.
6. ALWAYS COME TO CLASS IF AT ALL POSSIBLE! When you are absent, there is no way to fully make up for the class instruction you miss.
7. Always seek to understand, rather than simply to "squeak by." The grade you receive is important, but not nearly as important as the mental growth you gain from the process of learning the subject.
8. Memorization will help you absorb and retain factual information upon which understanding and critical thought is based. Knowing and using mathematical vocabulary and notation is the key to the understanding of the mathematical sciences.
9. PREPARE FOR TESTS! Your tests are often made up of questions that come directly from homework exercises, class notes, the chapter test, or the chapter review. Meet in study groups to discuss items that you think will be on the test. Use the study group for remediation and peer tutoring. Individuals who help others learn gain a better understanding themselves.
10. If you are making a serious effort and still not doing well, come in after school and talk with the teacher. He or she can probably help you overcome your difficulties.

Unsuccessful students who have disengaged from the classroom must be slowly integrated into the mainstream classroom to avoid possible embarrassment from their friends. The first time the student that plays the role of "sleeper" in the back of the classroom answers a question correctly, well-meaning friends of that student might laugh or make some sounds of surprise or astonishment that result in that student feeling ridiculed. That embarrassment could cause that student to revert to the defense mechanism that has served him so well—napping in the back row.

Since reluctant learners often sit in the back of the class, teachers might try moving them toward the front of the class in a very inconspicuous way—like making a new seating chart for the entire class. It would be wise to let that student know what you are doing and why you are doing it so they are not caught off guard making an inappropriate remark. All students should be encouraged to take notes, thereby participating in class.

The teacher could initially check for the reluctant student's understanding by simply asking if the lesson is understood (directive questioning)—not looking for verification by answering an academic conceptual question. As time progresses, the teacher might ask if the reluctant learner agrees with another student's understanding of the lesson. And finally, as the reluctant learner becomes more comfortable and immersed in the class, the classroom teacher should begin to ask the student questions that the student is likely to know and answer correctly—building success on success and self-confidence.

Building confidence and trust takes time. Unsuccessful learners have developed defense mechanisms over time to protect themselves from embarrassment and ridicule. Change is not usually embraced quickly or wholeheartedly by everyone—not even by classroom teachers or administrators. Realizing this, teachers need to work patiently with students. If teachers move too quickly, they may inadvertently embarrass the student they are trying to help. That embarrassment could result in a student's trust and confidence being shaken or broken in that teacher.

The reason students generally become reluctant learners is because they were unsuccessful learners. And the reason they were unsuccessful is probably because they didn't understand what exactly was expected of them.

In the list of student suggestions above, for instance, it was suggested the students neatly copy any problems on the board into their notebook. While that is great, what if the student doesn't know how to take notes, doesn't title what is in the notes, doesn't date it, doesn't draw pictures, write definitions, look for patterns, or write procedures with examples?

Chances are, there are many students in a classroom who could use guidance in the configuration of their notebook. Suggesting something simple like leaving white space so there is no visual overload might help students study more effectively and efficiently. Your least successful students probably have nothing to take home to help them with their homework or prepare for tests. Students need to be taught how to take notes.

Another suggestion requires students to memorize information. Is it possible that unsuccessful learners don't know how to memorize material effectively or efficiently? Earlier, we discussed student learning. Telling a student to go home and study does not help that student if the child really doesn't grasp what he or she needs to do at home to study.

The list of suggestions is meaningless unless students know what they really mean and know how to apply them. Explaining those suggestions is part of teaching.

Many unsuccessful learners come from homes where the parents might not be able to help their children with schoolwork because they were unsuccessful learners themselves. Additionally, a disproportionate number of unsuccessful learners come from poverty, and their parents are more likely to work in the evening and not be available to assist their children with homework or to make suggestions on how to study effectively. All too often, these students are left to make decisions at very early ages—decisions about what to make for dinner and whether to watch television, study, or interact with their friends. While many of these students might have good intentions going out the schoolhouse door, when they get home, those intentions are not carried through. It is a lot easier to watch their favorite television show or talk to their friends than it is to stop having fun and study.

Because of this, it is important that teachers implement the teacher expectancies and the Components of an Effective Lesson. The components and expectancies provide structure to daily instruction that is helpful to students who have not experienced a great deal of success in a math class. The structure includes long- and short-term reviews, oral recitation that embeds information into short-term memory, and note-taking that helps students complete their daily homework assignments and prepare for tests, as well as the guided practice to monitor student learning.

While homework and home study are important, teachers need to use their class time effectively so that students learn as much as possible in class. Learning difficulties among special populations stem largely from instructional practices that do not build upon informal knowledge and do not foster learning or from teachers that do not monitor student learning. Special populations will experience difficulty if the instruction begins with the abstract and moves too quickly or if the instruction relies on memorizing mathematics by rote.

Thinking about what causes learning difficulties for special populations, one would realize those are the same factors that cause difficulty for the general student population as well. Good teaching matters!

BELIEF SYSTEMS

Listening to some teachers, one might conclude that they don't believe their students can learn. If that's the case, they need to resign and allow someone else to get the job done. Administrators and teachers must believe their students can succeed, that if they do a better job teaching and their students work harder, success will follow.

If you are really interested in your students succeeding, then you should *build success on success*. I have always used the first unit of the year as the unit I shape beliefs and teach students to study effectively and efficiently, as well as teach mathematics.

To build success on success, students must first experience success. So, overteach the first unit. The students overlearn it, all the while teaching them what kind of learners they are, their concentration times, how to take notes, how to study effectively and efficiently. Provide examples of how you remember important information, allow time at the end of the class for note review, ensuring they have the information they need to successfully complete their homework or prepare for a test.

My belief is that math can be taught to anyone willing to learn. Get the students to be successful on the first test and show them that success was based on what they did to prepare—not just on being smart—and they will be on their way to a great school year. Preparing them to learn will help them succeed and make you feel better about your students' accomplishments.

If you hear yourself or others talking about *those* students, not *my* or *our* students, then chances are you are not taking ownership in their success.

There is a disconnect, a disassociation that acts as a disclaimer to your part in your students' learning.

Elementary administrators have an advantage over secondary principals in that they have typically taught all of the subjects the teachers they supervise teach. They know the subject matter and are familiar with the sequencing, benchmarks, and instructional strategies to help students learn.

Secondary school principals normally come from subject-specific areas. They have backgrounds in math, social studies, physical education, or science. All too often, those who don't have a background in the natural sciences feel threatened by their secondary math teachers. Some will acknowledge they didn't understand math, they didn't get the "math gene," and that's why they were not successful in math. So when they evaluate their math teachers, they are looking at classroom environment, instructional strategies, classroom management, and not really paying close attention to the math content being delivered to the students. That has to change. A lesson's worth should be determined by what students learned—not how well the class seemed to go.

Administrators must also change their belief systems. Many administrators will sit in a math class and evaluate the instruction, knowing full well they did not understand the day's lesson. My guess is that if the administrator—who probably took the class in high school, graduated high school, earned a bachelor's and advanced degrees, and is mature—did not understand the lesson, then how would they expect a thirteen-, fourteen-, or fifteen-year-old to understand it? If administrators did not understand the lesson, they need to address that with their teachers, because it is doubtful that the students are getting it.

CONTRADICTORY RULES—MISUNDERSTOOD BEHAVIORS

As classroom teachers, there are times when we just scratch our heads in bewilderment because of student behavior. It seems that some students just find trouble, then continue to dig deeper when confronted about their behavior. That additional digging is often interpreted as a sign of disrespect by many teachers, but it might not be.

For instance, as a sign of respect, many cultures require eye contact when correction in behavior is being discussed. Other cultures, as a sign of respect, discourage eye contact and in fact the person on the receiving end of the message is taught to look downward. If classroom teachers are not sensitive to their students' backgrounds, they may interpret a student looking down as a sign of disrespect, which in turn creates an additional problem for the student to deal with.

In more affluent areas, when adults have disagreements, frequently they are handled through litigation. This method to resolve problems has been adopted in schools using peer mediation. Since a form of peer mediation is used at home to resolve conflicts, some students view this as very consistent with home life.

In less affluent areas, on the other hand, disagreements are often handled by fighting. The people who win those fights are often held in high regard. Students coming to school from those communities are taught at home to take care of themselves by fighting, and peer mediation may be construed as a coward's way to resolve a problem. The end result of this might be a student beating up another student to address a grievance—which is seen by educators as bad behavior.

These same types of misunderstandings happen quite often in other circumstances, and again the behaviors can be traced back to what is learned at home. For instance, when parents argue at home, children from more affluent homes might be more inclined to listen quietly or take leave. In less affluent areas, children try to diffuse volatile situations at home by using humor to decrease the tension. Learning from that home experience, students might try to apply it at school. For instance, teachers often have to correct the behavior of students. In some cases, rather than being a quick direction, the teacher might spend some time discussing the seriousness of the situation. If students coming from less affluent neighborhoods see this discussion escalating into a more volatile situation, they may say something that to them seems to be funny—not as a sign of disrespect, but as a way to calm the situation.

Again, if classroom teachers and administrators are not sensitive to their students' backgrounds, those students might find their way to the office for being disrespectful and possibly be punished for using a strategy they learned at home to successfully diffuse an uncomfortable situation.

POINTS TO REMEMBER

The reason that students are not held to adult standards is because they are not adults. They make mistakes in judgment because of their own experiences or lack of experiences. As the adult role model, classroom teachers must make every effort to ensure that their students know how to act in different situations. Without explicit guidance from teachers and school administrators, students may apply the rules and behaviors they learned at home to school—not realizing those rules contradict standards of behavior at school.

The importance of developing positive student-teacher relationships cannot be overstated, especially with students who don't care for school or who have not experienced much success in the classroom. The research strongly suggests that students will work and work harder for teachers out of loyalty.

Creating the expectation that all students will succeed, the belief that students will succeed, and developing attitudes that success is attainable will result in a successful year. Let's do it!

9

NOW WHAT? NEXT STEPS

Too many of us are like the steel ball in a pinball machine, deciding by default. We shoot out, bump into the obstacles, make a lot of noise, cause lights to flicker, and then go down the proverbial tube. People with a well thought-out plan play a lot longer with a great deal more success and enjoyment.

THE DEPARTMENT IMPROVEMENT PLAN

As teachers and administrators develop a plan, it should not be perceived as a five-year plan. The department plan must have immediate as well as long-term impact. That plan should incorporate the My Kid and commonsense standards as well as implementing the building-success-on-success model as a cornerstone.

The following topics should be identified as *teacher expectancies* adopted and reinforced by each member of the department and included in the department improvement plan.

1. First test: overteach, overlearn; teach content while teaching students how they learn, concentration times, and how to study.
2. Improve student-teacher relationships. Talk to students. Be positive!
3. Use linkage to introduce new or more abstract concepts and skills, develop concepts, and teach the big idea.

4. Use simple, straightforward examples to clarify concepts being taught when introducing new material. Don't bog students down in arithmetic.
5. Adopt a balanced approach to instruction, emphasizing vocabulary and notation, concept development and linkage, memorization of important facts and procedures, appropriate use of technology, and problem solving.
6. Fully implement the Components of an Effective Lesson (CELs) and teacher expectancies.
7. Adopt a homework format that includes what teachers' value, not just problem sets.
8. Include reading and writing in the instructional plans.
9. Test what you value, use the test template, and provide practice tests halfway through the unit to help students prepare for the real test. Use the more formal language students will see on high-stakes tests.
10. Require students to take notes, and use oral recitation to embed information in short-term memory.
11. Use reviews at the end of each class to address mastery and deficiencies and to prepare for high-stakes tests.
12. Use time effectively. Start the class on time, and end the class at the end of the period.
13. Use the star system on notes, homework, and practice tests.
14. After completing a unit, before the exam, review the concepts and skills. Compare and contrast problems so students can differentiate between them and choose the best methods to be used.
15. Review each question on the practice test so students know exactly what is expected of them.

EFFECTIVE SCHOOLS

School administrators cannot be left out of any school improvement effort. While we are describing a department improvement plan, educational research suggests that effective schools generally have strong instructional leaders, a safe and orderly climate, a school-wide emphasis on basic skills, high teacher expectations, and regular assessment of student progress. For that reason, the school administration has to be included in any plan.

Principals serving in areas of poverty and low achievement have a much greater impact on student achievement than principals in more affluent regions of the country. Most principals understand the need to build a culture of success, and they know that high turnover is detrimental to the success of their students. Successful principals go out and find the best teachers and do whatever they can to retain them.

According to the U.S. Department of Education, schools with high student achievement and morale show the following traits:

- Vigorous instructional leadership
- A principal who makes clear, consistent, and fair decisions
- An emphasis on discipline and a safe orderly environment
- Instructional practices that focus on basic skills and academic achievement
- Collegiality among teachers in support of student achievement
- Teachers with high expectations of their students
- Teachers who believe that their students, through hard work, can and will learn
- Frequent review of student progress

Effective schools have effective principals—principals willing to observe, supervise, and evaluate their classroom teachers—not just overseers.

TEACHER SUPERVISION

Teachers welcome instructional suggestions that result in increased student achievement, but they rarely receive them. According to the Department of Education, supervision that strengthens instruction and teachers' morale has the following elements:

- Agreement between supervisor and teacher on specific skills and practices that characterize effective teaching

- Frequent observation by the supervisor to see if the teacher is using these skills and practices
- A meeting between the supervisor and teacher to discuss the supervisor's impressions
- Agreement between the supervisor and teacher on areas for improvement
- A specific plan for improvement, jointly constructed by the teacher and supervisor

Many secondary teachers report that principal involvement in their classroom does not occur often. When principals do observe classroom instruction, those teachers say they rarely receive recommendations that are specific enough to implement in their classrooms that have any impact on instruction or increased student achievement.

For schools to improve performance, building-level administrators must be willing to *inspect* what they *expect* of their teachers. The research suggests that more than 75 percent of teachers surveyed indicate that the suggestions, recommendations, or directions given to them by their supervisors have little or no impact on classroom instruction. Many teachers report that their postobservation conferences were little more than signing the evaluation document.

More recent research seems to suggest that even when school administrators know a teacher is performing poorly, over 65 percent don't confront the situation straight on. The teacher unions seem to take a great deal of the blame for the poor performance of teachers. These research findings might suggest that school administrators are dropping the ball by not doing their job more effectively.

The recommendations made in this text are very easy to observe and monitor. Using the planning guide from chapter 2, principals can be involved and monitor how teachers are preparing their instruction and assessment. The recommendations based on the CELs introduced in chapter 1 are easily monitored as well.

Whether the day's objective is on the board or whether the lesson is closed by restating the objective and providing a brief overview, having students write about what they learned at the close of the lesson, can be easily observed and monitored by the principal. Are teachers providing two review periods, one in the beginning of the period to go over recently taught material and the second to review long-term knowledge and prepare for high-stakes tests? Principals can easily determine if homework is more than just a problem set from a textbook and monitor whether students had an opportunity to practice with guidance from their teachers.

If the components and the teacher expectancies were adopted within a department improvement plan, principals could focus their recommendations on their implementation. While a checklist could be developed for the components, teachers would be much better served if the principal sat down with the teacher to discuss their observations in greater detail.

Although I said this before, it is worth repeating: Among school administrators, some will acknowledge they didn't understand math and weren't successful in math. So when they evaluate their math teachers, they tend to look more at classroom environment, instructional strategies, a checklist for the components, and classroom management, not really paying close attention to the math content being delivered to the students. That has to change. A lesson's worth should be determined by what students learned—not how well the class seemed to go.

Administrators must also change their belief systems. Many administrators will sit in a math class and evaluate the instruction, knowing full well they did not understand the day's lesson. If administrators did not understand the lesson, they need to address that with their teachers, because it is doubtful that the students are getting it.

Improvement plans rarely work unless the school's administration is an integral part of the plan and is actively participating in the process. Here are some suggestions to make this happen:

1. School administrators should meet with teachers by grade, by subject, or individually to explicitly go over teacher expectancies and how they will be evaluated using the CEL and teacher expectancies in the very beginning of the school year.
2. School administrators should confirm that classroom teachers are teaching their assigned curriculum from the very first day of instruction.
3. If school administrators did not understand the lesson they observed, then they need to discuss that with the teacher in the postobservation conference so the teacher can improve his or her instruction.

4. School administrators should observe the same classes, then meet and discuss their observations to ensure consistency in the evaluation and supervision of the teaching staff at their respective schools.
5. Classroom observations should begin in September or October with follow-up conferences scheduled within a week to provide specific suggestions, recommendations, or directions to improve instruction that will result in increased student achievement.
6. School administrators should examine teacher-made tests in common areas to determine whether they cover the same material with the same rigor in the same approximate time frames.
7. School administrators should require teachers to develop expected student grade distributions for each of their classes at the beginning of the school year.
8. School administrators should collect grade distributions on the very first test of the year in September, as an early warning sign to gain insight on where students or teachers need assistance.
9. School administrators should address the student population yearly to very explicitly go over academic and behavioral expectations.
10. School administrators should immediately address students not coming to school prepared with books, notebooks, pencils, or paper.
11. School administrators should support teachers. If students are observed off task in a classroom, administrators should go into that class and address students not engaged in learning.
12. School administrators should evaluate the effectiveness of a teacher based on student performance.
13. School administrators should provide timely feedback to teachers based on their observations.
14. School administrators should provide explicit suggestions, recommendations, or directions to teachers that will improve instruction resulting in increased student achievement.

EXPECT WHAT YOU INSPECT—INSTRUCTION MATTERS!

School administrators must observe classroom instruction early and often and have the belief that if they cannot follow or understand the instruction, then the student sitting beside them is probably not following it or understanding it either. Corrective action needs to be taken in the form of suggestions, recommendations, or directions.

On the following pages, I have provided an observation sheet that teachers and administrators might discuss and come to an agreement on regarding items that should be observable on most days during a regular class period. During the actual observation, the initial guiding questions can be used along with a time-coded description of what is occurring in the class. It is strongly recommended an observation take place over a two-day period so follow-up of discussions, explanations, and homework can be observed.

POSTOBSERVATION PROTOCOL

1. Go over the time-coded notes taken during observation and fill in answers in the Guiding Questions template.
2. Review grade distribution, performance on high-stakes tests (district or state), and benchmarks. Take into account the student population before coming to any conclusions about teacher effectiveness.
3. Determine the evidence you observed to justify the responses in the Guiding Questions template.
4. Use the responses in the Guiding Questions template to determine talking points for the postobservation conference.
5. To begin the discussion with a teacher, ask the teacher to identify areas that, in his or her opinion, went well. If there is agreement, then positively reinforce those and cross them off the list to be discussed. Be sure to include in your remarks specifically why they are being commended. For instance, "I liked the homework assignment because, before you assigned reading, you explicitly taught the vocabulary, connected the reading to prior learning, and previewed what they would read, and the next day you checked their understanding of what was read and corrected their understanding. Also, writing was assigned as part of the homework that included explanations, procedures, or formulas students need to know as well as exercises."

Guiding Questions
Addressing Instruction & Student Achievement

Class ____________________________________ Period______ Date__________

Teacher___________________________________

- How did the teacher prepare for the lesson? (test blueprint, examples, practice test, national tests, etc.)

- How did the teacher overtly plan for building success on success?

- How did the teacher plan to and implement positive student–teacher relationships?

---

1. Did the teacher employ a quick, crisp, purposeful *review* to prepare the students for the day's instruction?

2. Did the teacher provide an *introduction* that would lead to interest/enthusiasm in the day's instruction?

3. How was the *daily objective* communicated to the students?

4. Were you able to follow/understand the *instruction* through the concept and pattern development, or linkage in and out of discipline?

5. What types of opportunities were students provided to *practice*? (paced guided, group, independent)

6. How did the *homework* support the instruction? (vocabulary, notation, identifications, procedures, reading & writing)

7. How did the teacher *close* the lesson?

8. If time permitted, how did the teacher *review* and reinforce topics or address student deficiencies?

9. What evidence did you observe that *simple, straightforward examples* provided to students were chosen in advance to clarify instruction?

10. Did student *notes* support and reflect instruction, and could they be used to help study more effectively and efficiently? (pacing)

11. How was *vocabulary & notation* introduced and emphasized?
12. How was *reading/writing* incorporated in the student learning?
13. What methodologies did the teacher employ to assist students *memorizing* important information?
14. What types of *questioning strategies* did the teacher employ to ensure student understanding? (directed, highlighting, echoing, cueing, conceptual)
15. Was the technology used appropriate? (use of board)
16. How were students provided opportunities to *apply* their knowledge?
17. How does the teacher *prepare* students for a *unit test* to address student success?

- How were *assessments designed* to support student success? (blueprint, parallel constructed, grade distribution, timely feedback)

---

6. Ask the teachers to identify areas they think might be improved upon. If you are in agreement, especially with their priority, then determine the best way to have these addressed. If teacher acknowledges and accepts the recommendation, you can provide an oral recommendation or write those as suggestions, recommendations, or directions in their evaluations.
7. If the teacher does not identify any areas that can be improved upon, ask, "Is this the best you can do?" If they conclude they cannot grow any further by improving instruction or assessment, then you will need to be more directive.
8. Identify areas of concern that were not self-identified by teacher, discuss your observation and possible remedies, come to an agreement or understanding, then write these as recommendations or directions, based on teacher acceptance and acknowledgment of the concern.
9. Close the conference by restating the expectations that were identified at the beginning of the school year and the suggestions, recommendations, and/or directions that are related to the latest observation.
10. After the conference, write the evaluation and have the teacher come back to read and sign it.

In the Guiding Questions document, items 10 and 13 ask about notes and methodologies, respectively. A weakness that has been noted in many classes deals with procedural fluency. The students are learning to do exercises by repeated examples without the assistance of clear and concise step-by-step procedures. Procedures should be developed as part of conceptual development or linkage, written on the board, copied in the notes, orally recited with a visual link to the exercise, and used to do practice problems, complete homework, and prepare for unit tests.

POINTS TO REMEMBER

Using data and giving classroom teachers timely feedback is important to the success of students. School administrators must provide clear expectations to their staff at the beginning of the school year and check on the implementation of those expectations. The quality of instructional leadership and the classroom teacher will determine the success of the students at a school.

School administrators must be willing to address poor instruction and provide suggestions, recommendations, or directions that have an impact on instructional and assessment strategies that will result in increased student achievement.

10

CLOSING— SUMMING IT UP

ORGANIZING STUDENT LEARNING

The good news is that the recommendations in this book mirror practices teachers generally employ every day. The difference between what is currently done in the classrooms and the recommendations in this book are the refinement of those practices that help organize and focus student learning, resulting in increased student achievement.

ISSUES THAT NEGATIVELY IMPACT STUDENT ACHIEVEMENT

As we introduce the new, more rigorous common core standards in mathematics to students already struggling and earning poor grades, we have to be mindful of past practices that have had minimal effect on student achievement and learn from those experiences.

Too often, school districts and schools buy a new program and think their job is done. We know that doesn't work, but it provides school leaders cover, an opportunity to tell their communities they are doing something. A program is only as good as the teacher using it. Don't confuse activity with achievement. *What works is work.*

One issue that continues to be a need to be addressed is school administrators not explicitly communicating expectations to teachers. The research about supervision that indicates that 65 percent of administrators don't directly confront poor instruction is troublesome. So is the fact that 76 percent of teachers surveyed suggest that the comments by their supervisors have little or no impact on improving instruction. This can and needs to be resolved immediately. You can expect only what you are willing to inspect.

Many school administrators seem more concerned with employing the latest programs with the most bells and whistles meant to impress the casual observer than going into the classroom to observe instruction. What should have greater importance and value is answering a few simple questions: Are the students learning? Are they more comfortable in their knowledge, understanding, and application of mathematics? Is that being reflected in student grades and achievement levels?

At the end of the day, while learning can be fun, it is still work. Being involved in a project may be impressive and can create interest and enthusiasm in learning mathematics. But students still need to know information—they still need to study. That means students still need to memorize information and have enough practice to become procedurally fluent. Like all jobs and professions, people still have to do their homework. Attorneys study and plan their cases; doctors need to continually upgrade their knowledge and skills. Yes, surgeons might tell you they look forward to operating, but they still need that book knowledge and practice before they work on patients.

Whether teachers tend to be constructivist or traditionalist, supervisors should be concerned about student learning, especially struggling students. And to assist that learning, students should be set up to succeed by connecting their preparation and instruction to notes, homework, test prep, and tests.

ESSENTIAL TIPS TO INCREASE STUDENT ACHIEVEMENT

Being an educator is not just a job, it's a great profession—a profession that impacts not only the lives of individual students but also the nation as a whole. While education should be taken seriously, that does not mean it should not be enjoyed. Smile, enjoy what you do, it is important!

Everything you just read will mean absolutely nothing and will not result in increased student achievement unless classroom teachers and administrators believe our students can be successful. That belief must be on full display; by what we say and how we say it, with facial expressions and body language.

For students struggling in math, students who regularly earn grades of D or F in math, this book addresses the fundamental question that should be asked of teachers and administrators: What are *you* doing to help *my* child learn math? The short answer is to have teachers implement the Components of an Effective Lesson with certain expectancies being emphasized.

To increase student achievement, students need to be taught how to study more effectively and efficiently. Many struggling students honestly don't know how to do this. Classroom teachers interested in accelerating student achievement should take a closer look at their own practices to ensure they are organizing their instruction in ways that help students learn.

The seven, or 6+1, recommendations in this book are strategies successful teachers employ to ensure instruction is designed to help students learn successfully:

1. Lesson preparation and planning
2. Lesson delivery that includes concept development and linkage
3. Student notes that reflect and reinforce that instruction
4. Homework that supports and reflects the notes and the instruction
5. Test preparation, using a practice test that was constructed before instruction began
6. A balanced assessment that reflects state standards and the school district curriculum and reflects and reinforces the instruction, notes, homework, and test preparation
7. The "+1" reflects the importance of student-teacher relationships to improving student achievement.

Making the connections between the instruction, student notes, homework assignments, test preparation, and assessments helps students focus and study more effectively and efficiently, resulting in increased student achievement.

Instruction matters! So that means planning matters, too. Teachers should have a very clear idea of what they expect students to learn before they present their lessons. This means that before teachers begin to teach a unit, they should create a specification sheet, assessment blueprint, and test that reflect their expectations and what they value in that unit. Good mathematics instruction requires that ideas are developed and translated to procedures, rules, postulates, theorems, and corollaries that students would be able to reconstruct over time.

What should typically *not* be happening in a classroom is having a rule placed on the board that seemingly came out of nowhere, then having four, five, or six problems provided as examples or student classwork. Evidence of preplanned instruction should include sample problems, a practice test, a specification sheet, or an assessment blueprint that reflects what teachers expect students to know, recognize, and be able to do after instruction.

Memory researchers have identified "writing it down" as the number-one memory aid they use to remember. In school, we identify "writing it down" as note taking. Middle school and high school students are typically taking six, seven, or eight classes per day. They are receiving an enormous amount of information daily, information that in many cases is not very well organized. Many classroom teachers believe that it is a student's responsibility to take notes—that they have no responsibility in that endeavor—and that's unfortunate.

If teachers cannot visualize what student notes should look like before they start to teach the lesson, then it is very doubtful student notes will be organized in a way that will promote student learning. Notes are important in assisting students to study more effectively and efficiently, resulting in increased student achievement. To ensure students achieve this, school administrators should examine student notes and look for the components that were discussed in chapter 4.

Many homework assignments in math are nothing more than page numbers and exercises. Homework should encourage studying and support learning. It should reflect and support instruction. A school administrator should look at daily homework assignments to ensure they support instruction and increase student performance in class. Homework recommendations that encourage study and help students become more successful were discussed in chapter 5. There, evidence of well thought-out homework assignments was made clear for supervisors of mathematics.

Using a practice test that was created before instruction began that reflects and supports instruction, notes, and homework will assist struggling students, students who don't typically experience success in a math class, in organizing their learning and will result in increased student achievement. The practice exam is another tangible that can be viewed by the school administrator to ensure students are learning how to learn and learning the content that is being delivered to them.

If the real test is balanced and is constructed in parallel with the practice test, struggling students will be able to walk into an exam situation more confidently because they know what they know and what is expected of them. Again, school administrators can and should examine these exams for curriculum, balance, fairness, and portability of grades. That practice test is the light at the end of the tunnel and can be used to motivate students.

There is a great deal more to effective teaching that results in increased student performance than standing and delivering. While preparation, planning, and high-quality instruction that is connected to student notes, homework, test preparation, and tests are essential, positive student–teacher relationships are also critically important for student buy-in. How that alone affects student performance cannot be overstated.

Just increasing student achievement is not enough. The teachers in the math department should sit down and ask themselves what the highest performing schools are doing. Some school has to be identified as the highest achieving—why not yours?

Without consciously thinking about being the best school, the chances of reaching that goal are zero. Schools should discuss what they need to do to become the highest performing school, a school where education is taken seriously, a model school. After identifying those traits and practices, a goal should be established, publicized, and constantly worked toward. Rather than having teachers and administrators coming to school every day working hard, they should be coming to school every day working to their identified goal.

This book has provided many suggestions and recommendations that can be implemented quickly to help organize and focus student learning so they can study more effectively and efficiently, at no cost, and that have the advantage of being readily observable when monitoring math instruction.

Finally, we must remember that when students don't remember information, the community associates that lack of memory with students not learning. As educators, we must employ practices in the classroom that increase memory. There is no one thing that will result in increased student achievement; there are many factors. You should not cherry-pick which of these recommendations to implement—they should *all* be implemented.

My experiences suggest that there is a major disconnect between teachers' expectations, what they teach in the classroom, the notes students take away from that instruction, the homework assignments given to support that learning, and the test preparation and tests used to monitor student learning that ultimately leads to a grade. The recommendations provided are very observable components of an effective lesson and teacher expectancies on which high-quality instruction is based.

And since the recommendations include practices teachers already use, making the modifications or refinements on how to prepare a unit, using linkage to introduce new topics, notes, homework, or test preparation should not be considered major changes. These recommendations reflect a very blue-collar, no-nonsense, commonsense approach to increasing student achievement for struggling students and students living in poverty that will give them the best chance at finding success in mathematics and in life.

It is said that death is the great equalizer. That is true, but for the living, the great equalizer is education. While education does not guarantee increased earnings over a lifetime, it often translates to just that. An increased income leads to a better lifestyle, which very well could lead to a happier life, not having to worry about paying for the basics of food, shelter, and insurance to take care of the health needs of a family.

POINTS TO REMEMBER

Six of the seven key recommendations are designed to organize student learning so students can study more effectively and efficiently. These recommendations are no-nonsense, commonsense strategies that suggest *what works is work*. By ensuring good preparation, good instruction will follow. And when preparation and instruction are supported and reflected in student notes, homework assignments, test preparation, and tests, students are more focused on what we expect them to know, recognize, and be able to do.

The seventh recommendation, the "+1," relates the importance of student–teacher relationships and the impact those relationships have on student achievement.

The idea of using linkage to introduce new concepts and skills was embedded in the recommendation regarding instruction. The idea of linking concepts and skills to previously learned math and outside experiences was woven throughout the text.

We have students in our classrooms for a year; the students will have to live the rest of their lives with the formal knowledge they take away from our instruction and the intangibles they picked up along the way. Our students only get one chance at receiving a high-quality education. We owe it to them to do our very best every day. Let's do it!

ABOUT THE AUTHOR

Bill Hanlon, director of the Southern Nevada Regional Professional Development Program, has been an educator for over thirty years. His educational experiences include teaching at the junior high, senior high, and college levels. He was the coordinator of Clark County School District's Math/Science Institute and was responsible for K–12 math audits. He served as vice president of the Nevada State Board of Education, regional director of the National Association of State Boards of Education, and as a member of the National Council for Accreditation of Teacher Education States Partnership Board. He was also a member of Nevada's standards writing team in mathematics. He hosted a television series "Algebra, you can do it!" and taught math at the University of Nevada, Las Vegas, to prospective school teachers. He is a speaker at conferences such as the National Association of Secondary School Principals, Association of Latino Administrator and Superintendents, National Council of Teachers of Mathematics, National School Boards Association, and Association for Supervision and Curriculum Development.

Printed in Poland
by Amazon Fulfillment
Poland Sp. z o.o., Wrocław